Minimizing and Exploiting Leakage in VLSI Design

Nikhil Jayakumar • Suganth Paul
Rajesh Garg • Kanupriya Gulati
Sunil P. Khatri

Minimizing and Exploiting Leakage in VLSI Design

Springer

Nikhil Jayakumar
Morse Avenue 1168
94089, Sunnyvale
USA
nikhil.jayakumar@gmail.com

Dr. Suganth Paul
5701 S. Mopac Expressway
Austin TX 78479
#1523
USA
suganthpaul@gmail.com

Dr. Rajesh Garg
6430 NE Alder St.
Hillsboro OR 97124
Apt. B
USA
rajeshgarg@tamu.edu

Dr. Kanupriya Gulati
311 Stasney St.
College Station TX 77840
Apt. 1205
USA
kgulati@tamu.edu

Dr. Sunil P. Khatri
Texas A & M University
Dept. Electrical & Computer Engineering
College Station TX
77843-3128
214 Zachry Engineering Center
USA
sunilkhatri@tamu.edu

ISBN 978-1-4419-0949-7 e-ISBN 978-1-4419-0950-3
DOI 10.1007/978-1-4419-0950-3
Springer New York Dordrecht Heidelberg London

Library of Congress Control Number: 2009939713

Printed on acid-free paper

Springer is part of Springer Science+Business Media (www.springer.com)

To our parents and our teachers

Foreword

Power consumption of Very Large Scale Integrated (VLSI) circuits has been growing at an alarmingly rapid rate. This increase in power consumption, coupled with the increasing demand for portable/hand-held electronics, has made power consumption a dominant concern in the design of VLSI circuits today. Traditionally dynamic (switching) power has dominated the total power consumption of VLSI circuits. However, due to process scaling trends, leakage power has now become a major component of the total power consumption in VLSI circuits. This book presents techniques to reduce leakage, as well as techniques to exploit leakage currents through the use of sub-threshold circuits.

This book consists of three parts. In the first part, techniques to reduce leakage are presented. These include an algebraic decision diagram (ADD) based approach to implicitly represent the leakage corresponding to all possible inputs to a combinational design, a heuristic technique to find the minimum leakage vector in the presence of random Process, Voltage and Temperature (PVT) variations using signal probabilities, a low-leakage ASIC design methodology that uses high-V_T sleep transistors selectively, a methodology that combines *input vector control* and circuit modification, and a scheme to find the optimum reverse body bias voltage to minimize leakage.

As the minimum feature size of VLSI fabrication processes continues to shrink with each successive process generation (along with the value of supply voltage and therefore the threshold voltage of the devices), leakage currents increase exponentially. Leakage currents are hence seen as a necessary evil in traditional VLSI design methodologies. We present an approach to turn this problem into an opportunity. In the second part of this book, we attempt to exploit leakage currents to perform computation. We use sub-threshold digital circuits and come up with ways to get around some of the pitfalls associated with sub-threshold circuit design. These include a technique that uses body biasing adaptively to compensate for PVT variations, a design approach that uses asynchronous micro-pipelined Network of Programmable Logic Arrays (NPLAs) to help improve the throughput of sub-threshold designs, and a method to find the optimum supply voltage that minimizes energy consumption in a circuit.

While the second part of the book goes into details of various sub-threshold design approaches, the third part of this book presents silicon validation of these approaches. The third part of this book presents design and implementation details of a sub-threshold wireless BFSK transmitter chip. This chip was designed and fabricated to prove the feasibility of the sub-threshold design approaches detailed in the second part of this book. We also present results from tests carried out on the fabricated die that prove the value of sub-threshold design.

This book will serve as a valuable reference to anyone interested in understanding leakage currents in modern day DSM processes and to those interested not just in leakage reduction but also in how to exploit it to make practical ultra-low power integrated circuits.

Sunnyvale, CA Nikhil Jayakumar
Austin, TX Suganth Paul
Portland, OR Rajesh Garg
College Station, TX Kanupriya Gulati
College Station, TX Sunil P. Khatri

Preface

Power consumption is a major concern in today's VLSI designs. In particular, leakage power has become a significant component of the total power consumption of a chip and has thus received much attention in recent Deep Sub-micron (DSM) processes.

This book consists of three parts. The first part of this book addresses leakage *reduction* approaches while the second explores techniques to *exploit* leakage currents to perform computation. In the third part of the book, we present a test application of the techniques presented in the second part.

Since leakage power consumption is seen as a major issue in VLSI design today, there has been significant research into techniques to reduce leakage. In Part I of this book, new techniques to reduce leakage are proposed. These include an algebraic decision diagram (ADD) based approach to implicitly represent the leakage corresponding to all possible inputs to a combinational design, a heuristic technique to find the minimum leakage vector in the presence of random Process, Voltage and Temperature (PVT) variations using signal probabilities, a design approach that uses high-V_T *sleep transistors* selectively, a technique that modifies a circuit to reduce leakage while simultaneously finding the best input vector that minimizes leakage and a scheme to find the optimum reverse body biasing voltage to minimize leakage.

In the second part of this book, we attempt to exploit leakage currents rather than minimize them. We propose the use of sub-threshold digital circuits and present ways to get around some of the pitfalls associated with sub-threshold circuit design. These include a self-adjusting adaptive body-biasing technique that helps make a sub-threshold circuit less sensitive to PVT variations, a design approach that helps improve the throughput of sub-threshold designs through the use of asynchronous micro-pipelined Network of Programmable Logic Arrays (NPLAs), and a method to find the optimum supply voltage that minimizes energy consumption in a circuit.

In the third part of this book, we go over design details of a sub-threshold wireless BFSK transmitter IC. Data gathered from experiments carried out on the fabricated die are also presented along with a comparison to regular standard-cell-based version of the BFSK circuit.

Book Outline

This book is organized into three parts.

Part I of the book focuses on minimizing leakage. In Chap. 2, we survey some existing approaches to leakage reduction. This chapter is a good starting point to anyone interested in knowing the basic set of tricks used by digital designers today to tackle the problem of leakage currents. ADD-based exact and approximate techniques to implicitly compute the leakage of a combinational design for all possible inputs are presented in Chap. 3. Chapter 4 describes a heuristic approach for computing the minimum leakage vector for a combinational circuit using signal probabilities. This approach is further extended to account for random PVT variations. In Chap. 5, we present a new low-leakage standard cell-based ASIC design methodology, called the "HL" methodology that achieves leakage reduction through selective use of low-leakage variants of a standard cell. In Chap. 6, another design approach is presented that reduces leakage through using different variants of a standard cell and "parking" the circuit in its lowest leakage state. In Chap. 7 some experimental results are presented to prove that there is an optimum reverse body bias voltage for leakage minimization and then details of a circuit that can find this optimum reverse body bias voltage are presented.

In Part II of this book, we look at leakage currents differently and present practical techniques and methodologies that exploit leakage to perform computation. In Chap. 9, the reader is introduced to the idea of operating circuits in the sub-threshold region and thus exploiting leakage. This is a useful chapter to anyone interested in understanding the basics of sub-threshold circuit design and operation. In Chap. 10, we present a sub-threshold design methodology that compensates for the high sensitivity of sub-threshold circuits to Process, Voltage and Temperature (PVT) variations. This is a recommended chapter for readers who design or are planning to design ultra-low power (low voltage) circuits apart from sub-threshold circuits; the methodology presented in this chapter can also be applied for circuits operating at extremely low voltages near the sub-threshold region of operation. In Chap. 11, we discuss how the optimum voltage for low energy can often be much higher than the optimum voltage for power. In Chap. 12, an asynchronous micropipelined design flow and methodology is presented to alleviate some of the speed concerns of sub-threshold circuits.

In Part III of this book, we present details of how we implemented a sub-threshold BFSK transmitter IC that utilizes some of the sub-threshold design techniques presented in Part II of this book. It is recommended that the reader read this part of the book only after reading Part II of this book (specifically Chap. 10). In Chap. 14, the architecture of the transmitter is explained in detail. Chapter 15 delves into the implementation details of the IC. Some results from the experiments performed on the fabricated die are presented in Chap. 16.

Sunnyvale, CA Nikhil Jayakumar
Austin, TX Suganth Paul
Portland, OR Rajesh Garg
College Station, TX Kanupriya Gulati
College Station, TX Sunil P. Khatri

Acknowledgments

This book contains the results of several years of research by its authors, starting in 2003. The work presented in this book has been possible – thanks to the support from many sources.

The contents of this book are the result of research first started by two of the authors (Dr. Nikhil Jayakumar and Dr. Sunil P. Khatri) at the University of Colorado at Boulder. We would like to thank the students and faculty at Boulder, where our research on leakage power was initiated. We also wish to thank the students and faculty at Texas A&M University, where we continued our research into leakage and published several more papers in the area.

The work presented in this book would not have been possible without the tremendous amount of help and encouragement we have received from our families, friends, and colleagues.

First we would like to gratefully acknowledge the funding support without which the subthreshold transmitter IC would not have been possible. This includes support from Lawrence Livermore National Laboratories (LLNL) and the National Center for MASINT Research (NCMR). The support of Drs. Sheila Vaidya and Pete Bythrow is especially appreciated.

The work presented in this book would not have been possible without the tremendous amount of help and encouragement we have received from our families, friends, and colleagues.

Contents

Abbreviations

ADD	Algebraic decision diagrams
ATPG	Automatic test pattern generation
ASIC	Application specific integrated circuit
BDD	Binary decision diagrams
BER	Bit error rate
BFSK	Binary frequency shift keying
BPSK	Binary phase shift keying
BPTM	Berkeley predictive technology model
BTBT	Band-to-band tunneling
CCR	Channel-connected region
CMOS	Complementary metal oxide semiconductor
DAC	Digital to analog converter
DFF	D flip-flop
DLL	Delay locked loop
DSM	Deep sub-micron
DTMOS	Dynamic threshold MOS
EDP	Energy delay product
ESD	Electrostatic discharge
FFT	Fast fourier transform
FPGA	Field programmable gate array
FSK	Frequency shift keying
GEDL	Gate edge drain leakage
GIDL	Gate induced drain leakage
HDL	Harware description language
IC	Integrated circuit
ILP	Integer linear programming
ITE	If-then-else
IVC	Input vector control
LCM	Leakage current monitor
LSB	Least significant bit
LUT	Lookup table
LVS	Layout versus schematic
MDD	Multiple-valued decision diagram

MLV	Minimal leakage vector
MSB	Most significant bit
MTBDD	Multi-terminal binary decision diagram
MTCMOS	Multiple threshold CMOS
NCO	Numerically controlled oscillator
NPLA	Network of programmable logic arrays
OBDD	Ordered binary decision diagram
PCA	Principal component analysis
PDP	Power-delay-product
PLA	Programmable logic arrays
PVT	Process, voltage and temperature
ROBDD	Reduced ordered binary decision diagrams
RTL	Register transfer language
RVA	Random vectors approach
SDR	Software defined radio
SFDR	Spurious free dynamic range
SNR	Signal to noise ratio
SPICE	Simulation program with integrated circuit emphasis
STA	Static timing analysis
VCDL	Voltage controlled delay line
VLSI	Very large scale integration
VTCMOS	Variable threshold CMOS

List of Tables

List of Figures

Chapter 1
Introduction

1.1 The Need for Low Power Design

Since the advent of CMOS technology, an increased number of transistors per die and greater performance have been the primary driving factors for the semiconductor industry and process technology. The ability to integrate more transistors per die allowed chip manufacturers to put more components of a system into a single package and thus reduce not only just the sizes of the electronic devices we use today but also the cost and delay. The intense competition in the semiconductor industry has forced chip manufacturers pursue these goals aggressively. To the credit of the semiconductor industry, these goals (more transistors per die and greater performance) have been growing at an exponential rate, following Moore's law. However, in the process, the power dissipation of the Integrated Circuit (IC) has been growing at an alarming rate as well. In recent times, the excessive power consumption of contemporary circuits has become a dominant design concern [2]. In fact, the issue of power dissipation is one of the main concerns that has hampered the further scaling of transistors. A Very Large Scale Integrated (VLSI) chip consists of many energy storage elements, mainly capacitors, some that are required for computation (MOSFET device capacitances) and some that are a hindrance to circuit operation (parasitic capacitances). These capacitors are continually charged and discharged through resistive elements during circuit operation, resulting in energy dissipation in the form of heat. The amount of heat dissipated puts a restriction on the computational performance of the circuit, or the number of times the transistors in the circuit can switch for a given power budget. One could argue that the shrinking of devices has reduced the amount of parasitic capacitance and this alleviates power dissipation problems. However, the increase in the *number* of devices due to the increase in device density has more than compensated for the decrease in the parasitic capacitance of a single device.

In addition to shortened battery life for portable electronics, higher power consumption results in aggravated on-chip temperatures, which can result in a reduced operating life for the IC [3].

For portable electronics, longer battery life is the most important design constraint. As a result, low power consumption becomes a crucial requirement for

N. Jayakumar et al., *Minimizing and Exploiting Leakage in VLSI Design*,
DOI 10.1007/978-1-4419-0950-3_1, © Springer Science+Business Media, LLC 2010

circuits used in portable electronics. In fact, the rapid growth in the demand for portable electronics is one of the major drivers that has forced semiconductor manufacturers to make conscious efforts to reduce power consumption.

However, power consumption is not an issue just for portable electronics today. ICs that consume more power also dissipate more heat and this necessitates more expensive cooling solutions. In fact, the use of liquid cooling in high-performance desktop computers is now fairly common (especially in the gamer's market). In the consumer market, saving even a few cents per part can translate into significant profits for a company. Hence, an IC that dissipates a lot of heat and thus requires an expensive cooling solution directly impacts the cost of a system using the IC. For organizations that employ large server farms, the cost of cooling the servers and the power consumption of the servers themselves are significant, especially in this day and age of rising energy costs.

Hence, low power consumption is a zero-order constraint for most ICs manufactured today. In fact, *higher performance-per-watt* is the new mantra for micro-processor chip manufacturers today.

1.2 Leakage and Its Contribution to IC Power Consumption

The power consumption of a VLSI chip is broadly classified into two – dynamic power and leakage power. Dynamic power is also often referred to as active power or switching power. This is the power consumed when a transistor switches, transferring charge. Since this charge transfer is required for any computation, this source of power dissipation is often considered a more useful or necessary source of power dissipation.

On the other hand, leakage power is considered a wasteful expenditure of power. Leakage power is the power consumed when a *turned-off* device leaks current. This source of power consumption is considered wasteful expense and is the dominant source of power dissipation in many portable electronic devices (such as cell-phones, PDAs, etc.) that spend most of their time in the *standby* state.

As can be seen from Fig. 1.1 [1], IC power consumption has been increasing rapidly as we move to new technology nodes. Interestingly, while both dynamic as well as leakage power have been increasing, the leakage power component has been growing at a significantly faster rate. The reason for this trend is explained below.

Consider the n-channel MOS (NMOS) device. An NMOS device has four terminals, the drain, gate, source and bulk, and it operates in one of three modes of conduction [4, 6], depending on the voltage of its terminals (V_d, V_g, V_s, V_b, respectively). In the equations that follow,
$V_{xy} = V_x - V_y$.

- Sub-threshold region :
  $I_{ds}^{sub} = \frac{W}{L} I_{D0} e^{\left(\frac{V_{gs} - V_T - V_{off}}{n v_t}\right)} [1 - e^{-\frac{V_{ds}}{v_t}}]$
  when $V_{gs} < V_T$

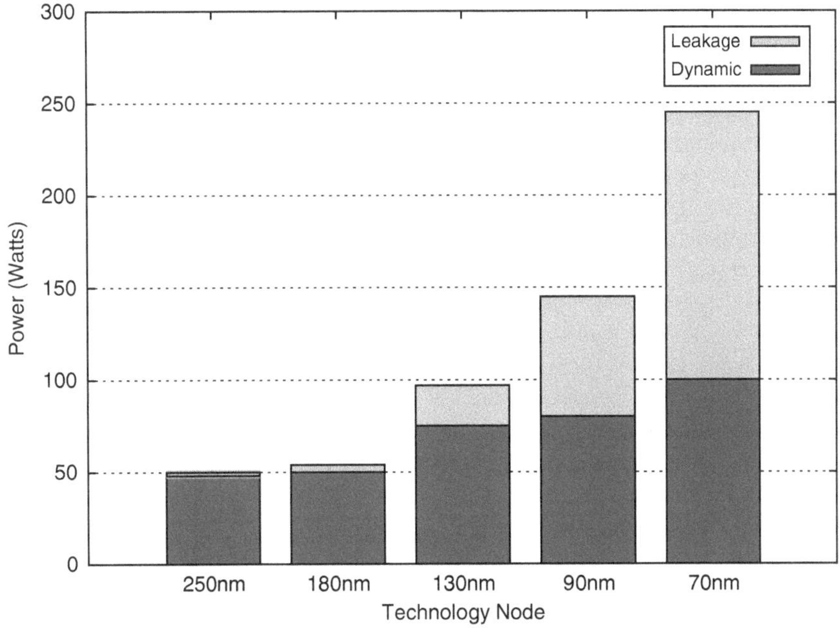

Fig. 1.1 Recent power trends [1]

- Linear (triode) region :
  $I_{\text{ds}}^{\text{lin}} = \beta\left([V_{\text{gs}} - V_{\text{T}}]V_{\text{ds}} - \frac{V_{\text{ds}}^2}{2}\right)$
  when $0 < V_{\text{ds}} < V_{\text{gs}} - V_{\text{T}}$
- Saturation region :
  $I_{\text{ds}}^{\text{sat}} = \frac{\beta}{2}(V_{\text{gs}} - V_{\text{T}})^2$
  when $0 < V_{\text{gs}} - V_{\text{T}} < V_{\text{ds}}$

The equations above express the current I_{ds} through an NMOS transistor in the three modes of conduction. In the above equations, V_{T} is the device *threshold voltage*. It depends on process-dependent factors like gate and insulator materials, thickness of insulator and channel doping density. It also depends on operational factors like V_{sb} (body effect)[1] and temperature (V_{T} is inversely proportional to device junction temperature). V_{T} is typically engineered to be about 20-25% of *VDD*. Also, $\beta = (\mu\varepsilon/t_{\text{ox}}) \cdot (W/L)$ where μ is the surface mobility of electrons (holes for a PMOS device) in the channel, ε[2] is the permittivity of the gate oxide, and t_{ox} is

[1] Body effect increases the threshold voltage of a device based on the following equation: $V_{\text{T}} = V_{\text{T}}^0 + \gamma\left(\sqrt{|(-2)\phi_{\text{F}} + V_{\text{sb}}|} - \sqrt{|2\phi_{\text{F}}|}\right)$, where V_{T}^0 is the threshold voltage at zero V_{sb}, γ is the body-effect coefficient – a physical parameter that expresses the impact of changes in V_{sb} and ϕ_{F} is the Fermi potential (typically 0.3 V for silicon).

[2] $\varepsilon = k \cdot \varepsilon_0$, where $k \equiv$ dielectric constant of the gate oxide.

the gate oxide thickness. W and L are the device width and length. Also, I_{D0} is a constant while $v_t = kT/q$. Here k is the Boltzmann's constant, q is the charge of an electron and $v_t = 26\,\text{mV}$ at room temperature. n is the sub-threshold swing parameter (a constant). Finally, V_{off} is a constant, typically equal to -0.08 V.

With technology scaling, supply voltages have been scaling down as well. The switching delay of a device is dictated by the current that can flow through it when the device is *turned on* (the device is in the saturation region). From the equation for the current of a device in the saturation region, it is clear that, to maintain a high saturation current and hence a small delay, any decrease in the supply voltage (which implies a decrease in V_{gs}) has to be accompanied by a decrease in the threshold voltage V_{T} of the device as well.

The leakage current for a PMOS or NMOS device corresponds to the I_{ds} of the device when the device is in the *cut-off* or *sub-threshold* region of operation. From the equation for I_{ds} in the sub-threshold region, we can see that the leakage current is exponentially dependent on the threshold voltage of the device. This is why a reduction in supply voltage (which is accompanied by a reduction in threshold voltage) results in exponential increase in leakage. Hence, with technology scaling and its accompanying supply voltage reduction, the leakage power consumption has been growing at a much faster rate than dynamic power consumption, as indicated in Fig. 1.1.

Another contributor to the greater rate of increase in leakage power is the fact that more logic is being integrated onto a single die. During operation however, there are only a few portions of the chip performing useful computations while a majority of the chip simply leaks, wasting power.

The power consumed by a design in the standby mode of operation is due to leakage currents in its devices. While the sub-threshold leakage current $I_{\text{ds}}^{\text{sub}}$ is the major component of leakage (in typical CMOS usage scenarios) there are several other sources of leakage as well. Figure 1.2 (adapted from [5]) shows the various sources of leakage for an NMOS device. In Fig. 1.2, I_{tox} represents the oxide tunneling current through the gate of the device, while $I_{\text{hot-e}}$ represents the gate leakage due to *hot-carriers* (electrons with high energy due to the applied electric field) being injected into the oxide layer of the gate. Gate leakage current is mainly due to these two components. The currents I_{pn} and I_{BTBT} are the currents that flow through the reverse-biased *pn* junction formed at the edges of the bulk and drain of the device. I_{pn} consists of mainly two components – a minority carrier diffusion/drift current and a current due to electron–hole pair generation. I_{BTBT} is the band-to-band tunneling (BTBT) current, which is a current due to the tunneling of electrons from the valence band of the p-region (from the bulk) to the conduction band of the n-region (to the drain). This tunneling happens due to a high electric field across the bulk–drain junction [which can happen when a Reverse Body Bias (RBB) is applied]. BTBT current is also referred to as bulk-BTBT or Gate Edge Drain Leakage (GEDL). I_{GIDL} is the Gate Induced Drain Leakage current (GIDL), which is also referred to as surface BTBT. This current occurs when the gate bias is negative relative to the drain. Under most operating scenarios and for most CMOS devices used today it is the sub-threshold leakage from the drain to the source of a device that dominates

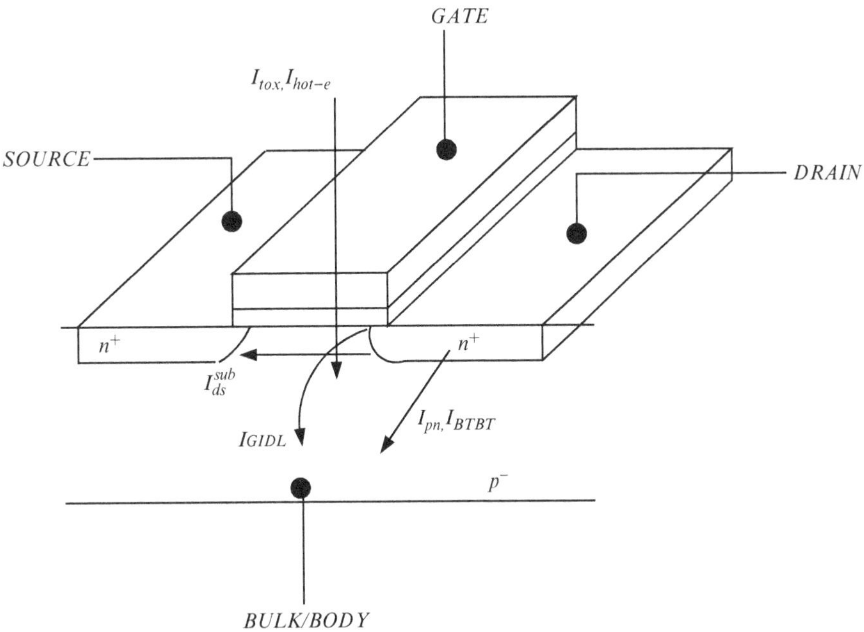

Fig. 1.2 Sources of leakage (NMOS device) (adapted from [5])

total leakage. In some situations (such as when there is a reverse body bias applied), the BTBT component may dominate. Because of process scaling trends (shrinking of gate oxide thickness) gate leakage has also become a concern. However, there is very little (apart from keeping supply and gate voltages low) that can be done at the design stage to tackle gate leakage. It is expected that the gate leakage issue would be tackled at the process technology stage.

With the prevalence of portable electronics, it is crucial to keep the leakage currents of a design small in order to ensure a long battery life in the standby mode of operation.

1.3 Summary

In this chapter, we have introduced the power consumption problem faced in VLSI design today. In particular, we have discussed why leakage power consumption is a major concern for today's designs. Starting with the next chapter, we discuss techniques to minimize leakage, followed by approaches to exploit leakage through the use of sub-threshold circuits.

References

1. Microprocessor Power Consumption. http://www.intel.com. Accessed on 5th May, 2005
2. The International Technology Roadmap for Semiconductors. http://public.itrs.net/ (2003). Accessed on 12th Nov, 2003
3. Daasch, W., Lim, C., Cai, G.: Design of VLSI CMOS Circuits Under Thermal Constraint. IEEE Transactions on Circuits and Systems II: Analog and Digital Signal Processing **49**(8), 589–593 (2002)
4. Rabaey, J.: Digital Integrated Circuits: A Design Perspective. Prentice Hall Electronics and VLSI Series. Prentice Hall, Upper Saddle River, NJ (1996)
5. Roy, K., Mukhopadhyay, S., Mahmoodi-Meimand, H.: Leakage Current Mechanisms and Leakage Reduction Techniques in Deep-Submicrometer CMOS Circuits. Proc. IEEE **91**(2), 305–327 (2003)
6. Weste, N., Eshraghian, K.: Principles of CMOS VLSI Design - A Systems Perspective, Addison-Wesley, Reading, MA (1988)

Part I
Leakage Reduction Techniques: Minimizing Leakage in Modern Day DSM Processes

In the first part of this book, we present some techniques and design methodologies aimed at minimizing leakage in digital integrated circuits. We first introduce some existing approaches to leakage reduction and then present some leakage reduction techniques invented by us.

1 Outline of Part I

Part I of this book is organized as follows. In Chap. 2, we discuss some previous leakage reduction approaches. In particular, we discuss Power-gating/MTCMOS techniques, Body biasing and Input Vector Control. The advantages and disadvantages of each of these techniques are also discussed in this chapter.

In Chap. 3, we describe an exact and approximate technique to compute the leakage current values for all input vectors in a combinational design. Apart from easing the task of finding the input vector that minimizes leakage, this technique also lets us plot a histogram of leakage values over all input vectors. This helps us evaluate different designs that may have similar minimum leakage currents for a particular input vector, but very different leakages for other input vectors seen during normal operation.

In Chap. 4, a heuristic to find a Minimal Leakage Vector (MLV) is presented. This heuristic uses signal probabilities at internal nodes to guide the search for the MLV. We also extend the heuristic to take statistical variation of leakage into account and find an optimal leakage vector that reduces the mean as well as the standard deviation of the leakage.

In Chap. 5, we describe a new low-leakage standard cell-based Application Specific Integrated Circuit (ASIC) design methodology that we call the "HL" methodology. This "HL" methodology is based on ensuring that during standby operation, the supply voltage is applied across more than one *off* device and there is at least one *off* device in the leakage path, which has a high V_T. For each standard cell in a library, we design two low-leakage variants. If the inputs of a cell during the standby mode of operation are such that the output has a high value, we use the variant that minimizes leakage in the pull-down network. Similarly we use the variant that minimizes leakage in the pull-up network if the output has a low value. While technology mapping a circuit, we determine the particular variant to utilize in each instance, so as to minimize the leakage of the final mapped design. We present experimental results that compare placed-and-routed area, leakage and delays of this new methodology against MTCMOS and a regular standard cell-based design style. The results show that our new methodology has better speed and area characteristics than MTCMOS implementations. The leakage current for HL designs can be dramatically lower than the worst-case leakage of MTCMOS-based designs and two orders of magnitude lower than the leakage of traditional standard cells. In contrast to the leakage of an MTCMOS design, the HL approach yields precisely estimable leakage values.

In Chap. 6, we present an approach that minimizes leakage by simultaneously modifying the circuit while deriving the input vector that minimizes leakage. This approach involves traversing a given circuit topologically from inputs to outputs and replacing gates to set as many gates as possible to their low-leakage state (in the sleep/standby state). The replacement does not necessarily reduce the leakage of the gate g being replaced, but helps set the gates in the transitive fanout of g to their low-leakage states. Gate replacement is performed in a slack-aware manner, to minimize the resulting delay penalty. One of the major advantages of this technique is that we achieve a significant reduction in leakage without increasing the delay of the circuit.

In Chap. 7, we first present results (from a 130 nm test chip) that prove that while sub-threshold leakage current decreases with applied Reverse Body Bias (RBB), another leakage component, the bulk Band-to-Band-Tunneling (BTBT) leakage component actually increases with applied RBB. We find that, there exists an optimum RBB that minimizes *total* leakage. We present a scheme that monitors the total leakage of a transistor and identifies the optimum RBB voltage that minimizes total leakage. Our method consists of a leakage current monitor and a digital block that senses the discharging (charging in the case of a PMOS transistor) of a representative *leaking* NMOS device in the design. Based on the speed of discharge, which is faster for leakier devices, an appropriate RBB value is applied. The scheme presented incurs very reasonable placed-and-routed area and power penalties in its operation.

Chapter 2
Existing Leakage Minimization Approaches

2.1 Leakage Minimization Approaches: An Overview

In recent times, leakage power reduction has received much attention in academia as well as industry. Several means of reducing leakage power have been proposed. Some of these are mentioned here.

2.1.1 Power Gating/MTCMOS

One of the natural techniques to reduce the leakage of a circuit is to gate the power supply using power-gating transistors (also called *sleep* transistors). Typically high-V_T power-gating transistors are placed between the power supplies and the logic gates. This is called the MTCMOS (Multi-threshold CMOS) approach [14, 17]. In standby, these power-gating transistors are turned off, thus shutting off power to the gates of the circuit. The MTCMOS approach can reduce circuit leakages by up to 2–3 orders of magnitude (depending on the threshold voltages and size of the sleep transistors used). However, the addition of sleep transistors causes an increase in the delay of the circuit. This delay penalty can be reduced by appropriately sizing up the sleep transistor. The downside to the up-sizing of the sleep transistor is the accompanied increase in the time and switching energy spent in *waking up* the circuit. As a consequence, power-gating (turning off the sleep transistors) is applied only when the circuit is expected to be in the standby state for a long period of time and when the wake-up time is tolerable. If a circuit using power-gating/sleep transistors goes in and comes out of the standby state too often, the power consumption may actually *increase* due to the higher power consumed in waking up the circuit. Another disadvantage of the MTCMOS approach is the fact that implementation of this technique requires circuit modification and possibly additional process steps (since high-V_T sleep transistors are used). Also, since cell inputs and outputs as well as bulk nodes float in an MTCMOS design operating in standby mode, the precise prediction or control of leakage is extremely difficult in MTCMOS. The voltage of these floating nodes can significantly affect the device threshold voltages. Hence, it is very difficult

N. Jayakumar et al., *Minimizing and Exploiting Leakage in VLSI Design*,
DOI 10.1007/978-1-4419-0950-3_2,

to precisely predict or control leakage in MTCMOS designs. Another drawback of MTCMOS is that memory elements in MTCMOS would require clean power supplies routed to them if we want to maintain their state in standby mode [17].

There has also been some research into the sizing of these sleep transistors. A conservative method to sizing the sleep transistors would be to first estimate the width of the sleep transistor required for each gate (or standard cell) in a design such that the delay of the individual gate is within a specified bound and then add up the sleep transistor widths for all gates to come up with the total sleep transistor width required. In [14], the authors propose a MTCMOS standby device sizing algorithm, which is based on mutually exclusive discharging of gates. This technique is hard to utilize for random logic circuits as opposed to the extremely regular circuits, which are used as illustrative examples in [14].

In [15], an MTCMOS-like leakage reduction approach was proposed, in which the MTCMOS sleep devices are connected in parallel with diodes. This ensures that the supply voltage across the logic is VDD $- 2V_D$, where V_D is the forward-biased voltage drop of a diode. The sub-threshold leakage current is significantly larger when $V_{ds} \gg nv_t$. This is because V_T drops due to the DIBL (Drain Induced Barrier Lowering) effect when V_{ds} is large [18]. The approach of [15] ensures that the V_{ds} across the sleep transistors is limited to VDD $- 2V_D$, thus keeping the sub-threshold leakage current low.

2.1.2 Body Biasing/VTCMOS

Increasing V_T via body effect and bulk voltage modulation is another way to reduce leakage power. The leakage current of a transistor decreases with greater applied Reverse Body Bias. Reverse Body Biasing affects V_T through body effect, and sub-threshold leakage has an exponential dependence on V_T as seen in the sub-threshold current equation (2.1).

$$I_{ds}^{sub} = \frac{W}{L} I_{D0} e^{\frac{V_{gs}-V_T-V_{off}}{n v_t}} \left[1 - e^{-\frac{V_{ds}}{v_t}}\right]. \quad (2.1)$$

The body effect equation can be written as:
$V_T = V_T^0 + \gamma(\sqrt{|(-2)\phi_F + V_{sb}|} - \sqrt{|2\phi_F|})$, where V_T^0 is the threshold voltage at zero V_{sb}, γ is the body-effect coefficient – a physical parameter that expresses the impact of changes in V_{sb} and ϕ_F is the Fermi potential (typically 0.3 V for silicon). Thus, the threshold voltage of devices can be dynamically adjusted using body biasing. Hence, this method of controlling the threshold voltage of transistors through body biasing is often referred to as the Variable Threshold CMOS or VTCMOS technology.

In [16], the authors describe how they applied VTCMOS technology to both the logic and memory elements of a 2-D Discrete Cosine Transform (DCT) core processor. During the active mode of operation, they apply a reverse body bias of 0.5 V and during standby they increase the reverse body bias to 3.3 V. The VTCMOS scheme implemented consisted of leakage current monitors (LCMs) to monitor the sub-threshold leakage and two charge-pump circuits – one to increase the applied RBB and another to decrease the applied RBB. These charge pumps were controlled in a closed-loop fashion using the leakage current monitors for feedback. In [12], the authors study the characteristics of VTCMOS for series connected circuits. They find that VTCMOS is effective for improving the performance of series connected devices too. In [11], the authors propose a compact analytical model of VTCMOS to help study the currents through a VTCMOS transistor during the active and standby states. They also study the influence of short channel effect (SCE) on the performance of VTCMOS.

The advantage with VTCMOS is that leakage current can be reduced in the standby mode by applying a reverse body bias (RBB) that raises the threshold voltage or the delay can be reduced in the active mode by applying a forward body bias that decreases the threshold voltage. However, with current technology scaling, the body-effect coefficient γ is reducing. Apart from this, there is also the overhead of implementing additional body-biasing supplies and the need to use special processes (such as the triple-well process) in order to provide separate well biasing. This method offers the advantage of decreasing the leakage in standby mode while not increasing the delay in the active mode.

In [4], the authors propose a dynamic threshold MOSFET design for low-leakage applications. In this scheme, the device gate is connected to the bulk, resulting in high-speed switching and low-leakage currents through body effect control. The drawback of this approach is that it is only applicable in situations where VDD is lower than the diode turn-on voltage. Also, the increased capacitance of the gate slows the device down, and as a result, the authors propose the use of this technique for partially depleted SOI (Silicon-On-Insulator) designs.

2.1.3 *Input Vector Control*

Another technique used to minimize leakage is the technique of *parking* a circuit in its minimum leakage state. This technique takes advantage of the fact that the leakage of a gate is dependent on the state of the inputs of the gate. The technique involves very little or no circuit modification and does not require additional power supplies. A combinational circuit is *parked* in a particular state by driving the primary inputs of the circuit to a particular value. In the standby mode, this value can be scanned in or forced using MUXes (with the standby/sleep signal used as a select signal for the MUX). This technique is frequently referred to as *input vector control (IVC)*. Finding the best (lowest leakage) input vector, also called the Minimal Leakage Vector (MLV) determination problem, is known to be an NP-hard problem. However, several heuristics have been developed to find an optimal vector. Researchers have used models and algorithms to estimate the nominal leakage current of a circuit [7, 8, 20]. In [10], the authors find a minimal leakage vector using random search with the number of vectors used for the random search selected to achieve a specified statistical confidence and tolerance. In [20], the authors reported a genetic algorithm-based approach to solve the problem. The authors of [13] introduce a concept called leakage observability, and based on this idea, they describe a greedy approach as well as an exact branch and bound search to find the maximum and minimum leakage bounds. The work of [9] is based on an Integer Linear Programming (ILP) formulation. It makes use of pseudo-Boolean functions, which are incorporated into an optimal ILP model and a heuristic mixed integer linear programming method as well. In [6], the authors present a Multiple-valued Decision Diagram (MDD) [5] based algorithm to determine the lowest leakage state of a circuit. The use of MDD-based MLV computations limits the applicability of [6] to small designs.

In [19], the authors present a greedy search-based heuristic, guided by node controllabilities and functional dependencies. The algorithm used in [19] involves finding the controllability and the controllability lists of all nodes in circuit and then using this information as a guide to choose gates to set to a low-leakage state. The controllability of a node is defined as the minimum number of inputs that have to be assigned to specific states in order to force the node to a particular state (based on concepts used in automatic test pattern generation) [2]. Controllability lists are defined as the minimum constraints necessary on the input vector to force a node to particular state. The time complexity of their algorithm is reported to ne $O(n^2)$, where n is the number of cells (gates) in the circuit. However, in estimating the complexity of their algorithm, it is not clear if the authors include the time taken to generate the controllabilities and controllability lists of each node in the circuit. While finding the controllabilities can be done fairly easily [2], generating the controllability lists can be more involved.

In [1,3], the authors express the problem of finding a minimum leakage vector as a satisfiability problem and use an incremental SAT solver to find the minimum and maximum leakage current. While their approach worked well for small circuits, the authors report very large runtimes for large circuits. The authors therefore suggest using their algorithm as a checker for the random search suggested in [10]. In [1], the authors introduced a method for controlling the internal nodes by modifying some gates, without using extra multiplexers. In addition, the delay constraints are explicitly accounted for and the optimal subset of internal nodes of the circuit to be controlled is determined by the SAT formulation.

2.2 Summary

In this chapter, we have presented some existing approaches to leakage power reduction. In the next few chapters, we propose some new approaches to tackle the leakage reduction problem.

References

1. Abdollahi, A., Fallah, F., Pedram, M.: Leakage Current Reduction in CMOS VLSI Circuits by Input Vector Control. IEEE Transactions on VLSI Systems **12**(2), 140–154 (2004)
2. Abramovici, M., Breuer, M.A., Friedman, A.D.: Digital Systems Testing and Testable Design. IEEE Press, New York, NY (1990)
3. Aloul, F., Hassoun, S., Sakallah, K., Blauuw, D.: Robust SAT-Based Search Algorithm for Leakage Power Reduction. In: Proc. Power and Timing Models and Simulation. Seville, Spain (2002)
4. Assaderaghi, F., Sinitsky, D., Parke, S.A., Bokor, J., Ko, P.K., Hu, C.: Dynamic Threshold-Voltage MOSFET (DTMOS) for Ultra-low Voltage VLSI. IEEE Transactions on Electron Devices **44**(3), 414–422 (1997)
5. Bahar, R.I., Frohm, E.A., Gaona, C.M., Hachtel, G.D., Macii, E., Pardo, A., Somenzi, F.: Algebraic Decision Diagrams and Their Applications. Formal Methods in Systems Design **10**(2/3), 171–206 (1997)
6. Chopra, K., Vrudhula, S.: Implicit Pseudo Boolean Enumeration Algorithms for Input Vector Control. In: Proc. Design Automation Conference, pp. 767–772. San Diego, CA (2004)
7. Duarte, D., Tsai, Y., Vijaykrishnan, N., Irwin, M.J.: Evaluating Run-Time Techniques for Leakage Power Reduction. In: 7th ASPDAC/15th International Conference on VLSI Design (2002)
8. Ferre, A., Figueras, J.: Characterization of Leakage Power in CMOS Technologies. In: Proc., IEEE International Conference on Electronics Circuits and Systems, pp. 85–188 (1998)
9. Gao, F., Hayes, J.: Exact and Heuristic Approaches to Input Vector Control for Leakage Power Reduction. In: Proc. International Conference on Computer-Aided Design, pp. 527–532. San Jose, CA (2004)
10. Halter, J., Najm, F.: A Gate-Level Leakage Power Reduction Method for Ultra Low Power CMOS Circuits. In: Proc. Custom Integrated Circuits Conference, pp. 475–478. Santa Clara, CA (1997)
11. Hyunsik, I., Inukai, T., Gomyo, H., Hiramoto, T., Sakurai, T.: VTCMOS Characteristics and Its Optimum Conditions Predicted by a Compact Analytical Model. In: Proc. International Symposium on Low Power Electronics and Design, pp. 123–128. Huntington Beach, CA (2001)
12. Inukai, T., Hiramoto, T., Sakurai, T.: Variable Threshold Voltage CMOS (VTCMOS) in Series Connected Circuits. In: Proc. International Symposium on Low Power Electronics and Design, pp. 201–206. Huntington Beach, CA (2001)
13. Johnson, M., Somasekhar, D., Roy, K.: Models and Algorithms for Bounds on Leakage in CMOS Circuits. IEEE Transactions on Computer-Aided Design of Integrated Circuits and Systems **18**(6), 714–725 (1999)
14. Kao, J.T., Chandrakasan, A.P.: Dual-Threshold Voltage Techniques for Low-Power Digital Circuits. IEEE Journal of Solid-State Circuits **35**(7), 1009–1018 (2000)
15. Kumagai, K., Iwaki, H., Yoshida, H., Suzuki, H., Yamada, T., Kurosawa, S.: A Novel Powering-down Scheme for Low Vt CMOS Circuits. In: Digest of Technical Papers, Symposium on VLSI Circuits, pp. 44–45. Honolulu, HI (1998)
16. Kuroda, T., Fujita, T., Mita, S., Nagamatsu, T., Yoshioka, S., Suzuki, K., Sano, F., Norishima, M., Murota, M., Kako, M., Kakumu, M.K.M., Sakurai, T.: A 0.9-V, 150-MHz, 10-mW, 4 mm 2, 2-D Discrete Cosine Transform Core Processor with Variable Threshold-Voltage (VT) Scheme. IEEE Journal of Solid-State Circuits **31**(11), 1770–1779 (1996)
17. Mutoh, S., Douseki, T., Matsuya, Y., Aoki, T., Shigematsu, S., Yamada, J.: 1-V Power Supply High-Speed Digital Circuit Technology with Multithreshold-Voltage CMOS. IEEE Journal of Solid-State Circuits **30**(8), 847–854 (1995)
18. Rabaey, J.: Digital Integrated Circuits: A Design Perspective. Prentice Hall Electronics and VLSI Series. Prentice Hall, Upper Saddle River, NJ (1996)

19. Rao, R., Liu, F., Burns, J., Brown, R.: A Heuristic to Determine Low Leakage Sleep State Vectors for CMOS Combinational Circuits. In: Proc. International Conference on Computer-aided Design, pp. 689–692. San Jose, CA (2003)
20. Zhanping, C., Johnson, M., Liqiong, W., Roy, W.: Estimation of Standby Leakage Power in CMOS Circuit Considering Accurate Modeling of Transistor Stacks. In: Proc. International Symposium on Low Power Electronics and Design, pp. 239–244. Monterey, CA (1998)

Chapter 3
Computing Leakage Current Distributions

3.1 Overview

With leakage power increasing as a fraction of the total power of a design, due to the current design trends, it is arguably important to find the leakage for all input vectors. This is useful when comparing candidate implementations of a design with the same minimum leakage values. An implementation that has a leakage histogram with larger number of input vectors contributing to lower leakage values would be preferred over other implementations. This would not only minimize the leakage during the regular operation of the circuit, but also ease the task of finding a vector that results in minimum leakage state.

The remainder of this chapter is organized as follows: The motivation for this work is discussed in Sect. 3.4. Some preliminary work necessary to understand the details of our approach is discussed in Sect. 3.3. Section 3.5 discusses previous work in this area. In Sect. 3.6 we describe our approach to compute leakage current distributions. We discuss the experimental results of our approach in Sect. 3.7. Conclusions and future work are discussed in Sect. 3.8.

3.2 Introduction

The approach described in this chapter is based on an Algebraic Decision Diagram (ADD) [3, 6] based computation, which enables the determination of the leakage values for all possible input vectors in the design. The approach is termed as AL^{all}. The exact version of AL^{all} is called AL^{all}_{ex}, while the approximate version is called AL^{all}_{app}. The determination the leakage values for all input vectors is useful in several contexts, such as the following:

- It allows the computation of the average, minimum and maximum leakage for the design in an accurate manner.
- It allows the construction the histogram of leakage values for a design. This can be of use when comparing two or more candidate implementations (with similar minimum or maximum leakage values) of a single circuit. The design with a

N. Jayakumar et al., *Minimizing and Exploiting Leakage in VLSI Design*,
DOI 10.1007/978-1-4419-0950-3_3, © Springer Science+Business Media, LLC 2010

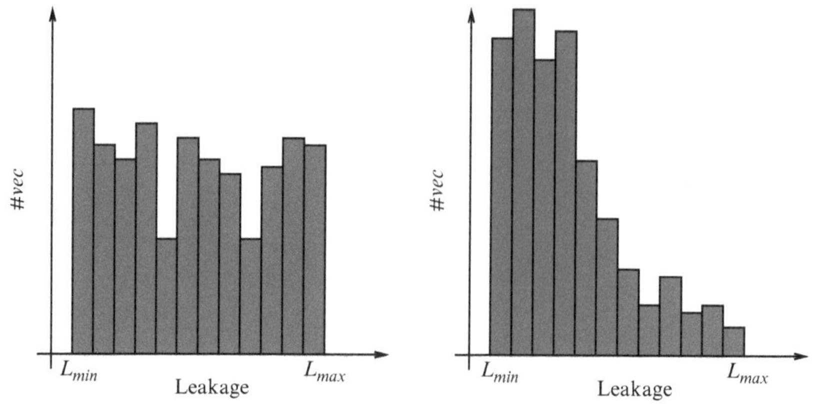

Fig. 3.1 Leakage histograms for two implementations of a design

leakage histogram that is skewed towards the lower leakage values would be preferred, since it would reduce dynamic power under normal operation. For example, during dynamic operation, the circuit may switch repeatedly between a set of vectors. In this case, the implementation that has a leakage histogram skewed towards lower leakage values would be preferred. Figure 3.1 illustrates this idea. The leakage histograms of two designs (with similar maximum leakage values) are shown. The histogram to the right is preferred, since it has a large number of vectors with low-leakage values.

- It enables the computation of the lowest leakage state for a design and the input vector corresponding to that state.

Clearly, an explicit representation of all leakage values would be infeasible. The problem of computing the leakage of all input vectors for a design is approached as follows. An Algebraic Decision Diagram (ADD) based approach is proposed to represent the leakage values of a circuit. The problem of building an ADD to implicitly represent the exact[1] leakage values of a design has been formulated and solved. In order to expand the applicability of this approach to larger designs, a method to implicitly compute the approximate leakage values of a design is also presented. These approaches can be used to construct the histogram of leakage values for a design. These data are beneficial when comparing two or more candidate implementations (with similar maximum leakage values) of a single circuit. Experimental data indicate that the approximate calculation of leakage values demonstrated a bounded loss of accuracy, with a significant improvement in the efficiency of the technique. Leakage histograms for area-mapped and delay-mapped versions of some benchmark circuits are computed, and their leakage characteristics are compared.

[1] The term *exact* used here and in the sequel refers to an *algorithmic* exact as opposed to an absolute exact.

3.3 Background

3.3.1 Reduced Ordered Binary Decision Diagrams

A reduced ordered binary decision diagrams (ROBDD) is a graphical representation of a Boolean function. It can represent many logic functions compactly as compared to a sum of product (SOP) or a truth table representation. Moreover, several logic operations like tautology checking and complementation can be performed on ROBDDs in constant time. For a particular variable ordering, an ROBDD is a canonical form of representing a Boolean function. However, it is more efficient in memory utilization than a truth table, which is another canonical representation of a Boolean function. As the name suggests, ROBDDs are a *reduced* form of BDDs with a particular *variable ordering*. The structure of the BDD and the reduction rules followed are described in the sequel.

A BDD represents a Boolean function as a directed acyclic graph (DAG), with each nonterminal node assigned to a variable of the function. It is also referred to as a Shannon cofactoring tree. Each node performs the Shannon cofactoring of the Boolean function represented by that node, with respect to the variable assigned to it. Figure 3.2 illustrates the BDD for the function $(x_1 + x_2) \cdot x_3$. Each node has two outgoing edges, corresponding to the positive cofactor of the node function with respect to the node variable (shown as a solid line) or the negative cofactor of the node function with respect to the node variable (shown as a dashed line). The terminal nodes (shown as boxes) are labeled with 0 or 1, corresponding to the possible function values. For any assignment to the function variables, the function value is determined by tracing a path from the root of the BDD to a terminal node following the appropriate positive or negative branch from each node. The number of vertices in the BDD is exponential in terms of the number of variables in the logic

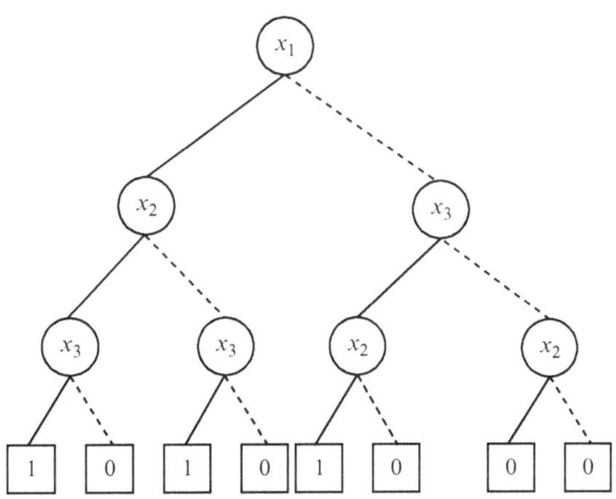

Fig. 3.2 Shannon cofactoring tree of logic function $(x_1 + x_2) \cdot x_3$

Fig. 3.3 OBDD of logic function $(x_1 + x_2) \cdot x_3$

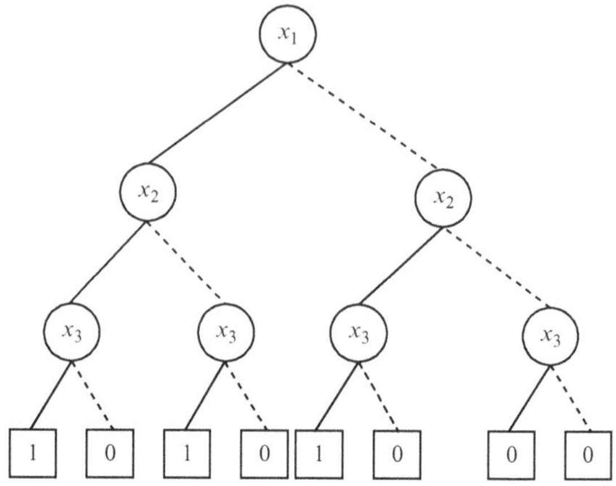

function. Therefore, for functions with a large number of variables, BDDs may not be a good choice for representing the function. In general, the variable ordering along different paths in the BDD can be different.

The graph in Fig. 3.2 is transformed into ordered BDDs (OBDDs) if we use a fixed variable ordering along any path from root to leaves. Consider the variable to be in the order $x_1 < x_2 < x_3$. That is, every path from the root to a leaf encounters variables in the order $x_1 < x_2 < x_3$. The resulting OBDD is shown in Fig. 3.3. In addition, on application of the following reduction rules on the OBDD, an ROBDD for the function is obtained.

- Remove nodes that have identical children.
- Merge nodes that have isomorphic BDDs.

ROBDDs are a canonical representation of a logic function for a given variable ordering. Figure 3.4 shows the resulting ROBDD when the above mentioned reduction rules are applied to the OBDD shown in Fig. 3.3. Note that even in an ROBDD, the number of nodes can be exponential in terms of the number of variables. The size of ROBDDs (i.e. number of nodes) depends upon the variable ordering. Therefore, variables must be ordered in a manner that minimizes the size of the ROBDD. Computing an optimum variable ordering is an NP-Complete problem. There are efficient heuristics available that can choose an appropriate ordering of variables, which results in the ROBDD of reasonable size. However, there are functions that have polynomial sized multi-level representations while their ROBDDs are exponential for all input orderings. A multiplier is an example of such a function. The terms ROBDD and BDD are used interchangeably in the rest of this chapter.

The following BDD operations are used in the approach presented:

- *bdd_find_minterm(f)*: This function returns one cube or minterm from all the existing cubes or paths to terminal node "1" of the BDD for f. This path is basically a cube in the onset of the Boolean function represented by f.
- *bdd_count_onset(f,var_array)*: This function counts the number of minterms in the onset of the function f, over the variables in *var_array* (single variable BDD

Fig. 3.4 ROBDD for logic function $(x_1 + x_2) \cdot x_3$

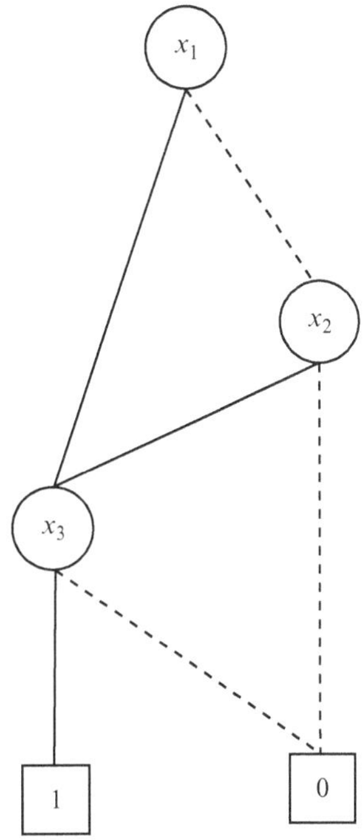

formulas). *var_array* must contain the variables in the support of f. For example, if $f = b \cdot d$, and *var_array* $= [a, b, c, d]$, then this function returns 4.

- *bdd_substitute(f, old_array, new_array)*: This function substitutes all variables from the array *old_array* with the corresponding variables from the array *new_array* in the BDD of "f." *old_array* and *new_array* are arrays of BDDs with equal cardinality. Given two arrays of variable BDDs a and b consisting of member values $(a_1 \,..\, a_n)$ and $(b_1 \,..\, b_n)$, this function replaces all occurrences of a_i by b_i in f. This operation is linear in the number of nodes in the BDD representation of f.

3.3.2 Algebraic Decision Diagrams

BDDs with multiple terminal nodes are called Multi-terminal BDDs (MTBDD). Because of their applicability to different algebras (including Boolean algebra) the term algebraic BDD was coined in [3]. A BDD can be viewed as an ADD with terminal values from the set {0,1}. An ADD with n terminals has terminal values selected from the set $\{ a_1, a_2, \cdots, a_n\}$, where a_i are algebraic or symbolic values. The values are also called *discriminants* of the ADD.
Some general properties of ADDs are as follows.

- ADDs are canonical. When dealing with ADDs with a large number of discriminants the usefulness of this property may decrease.
- Edge attributes such as complementation flags may be of limited utility, because complementation in Boolean algebra may not have a meaningful counterpart in the ADD context.
- These factors lead to a recombination efficiency (which arises due to sharing of isomorphic subgraphs), which is relatively small in comparison to BDDs.
- In comparison to other sparse data structures, ADDs provide a uniform $\log(N)$ access time where N is the number of real numbers being stored in the ADD.
- ADDs cannot beat sparse matrix data structures in terms of worst case space complexity. However, recombinations of isomorphic subgraphs may give considerable practical advantage to ADDs over other data structures.

An example of an ADD on three variables x_1, x_2 and x_3 is shown in Fig. 3.5. The discriminants here are not restricted to $\{0,1\}$. Also, note that the sharing mechanism is similar to that in a BDD, but since the terminal nodes can be of any numeric (or symbolic) value, the number of nodes shared could be fewer than those in a BDD.

The following ADD operations are used in the work presented:

- *ITE(f,g,h)*: The If-Then-Else (ITE) function takes three arguments. The first is an ADD restricted to have only 0 or 1 as terminal values. The second and third arguments are generic ADDs. *ITE* is defined as

$$ITE(f,g,h) = f \cdot g + f' \cdot h$$

  ITE can be applied as a recursive procedure for traversing through an entire ADD structure.
- *ADD_threshold(f,g)*: This function thresholds the discriminants of ADD f against a constant g. If the value of a terminal node is greater that or equal to g, it keeps the terminal node value as it is, else it assigns the terminal node to a value 0 or *FALSE*.

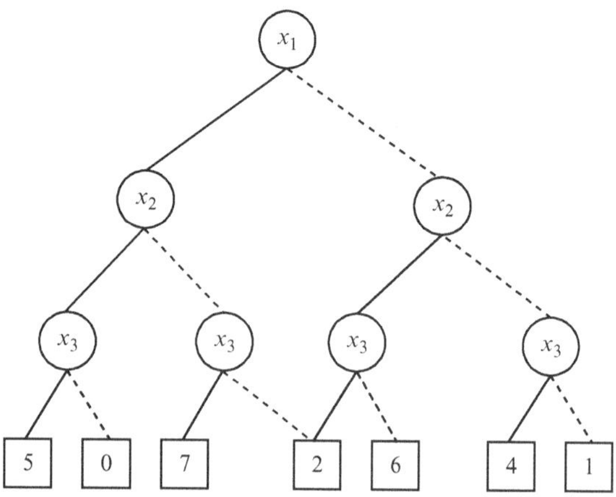

Fig. 3.5 An example ADD on three variables x_1, x_2, and x_3

- *ADD_to_BDD(f,t)*: This function is identical to *ADD_threshold(f,t)* except that when the value of a a terminal node is greater than or equal to t, the terminal node is assigned the value 1 or logical *TRUE*. In effect, the decision diagram is left with terminal nodes belonging to the set $\{0,1\}$ and hence is now a *BDD*.
- *cofactor(f,g)*: This function returns Shannon cofactor of an ADD f with respect to ADD g. g must be an ADD or a BDD of a cube.

3.4 The Intuition Behind Our Approach

Table 3.1 shows the leakage of a NAND3 gate for all possible input vectors to the gate. The leakage values shown are from a SPICE simulation using the 0.1-μ BPTM [4] models, with a VDD of 1.2 V.

As can be seen from Table 3.1, setting a gate in its minimal leakage state (000 in the case of the NAND3 gate) can reduce leakage by about 2 orders of magnitude. Ideally, it is desirable to set *every* gate in the circuit to its minimal leakage state. However, this may not be possible due to the logical inter-dependencies between the inputs of the gates. Finding this minimum leakage state as stated in Chap. 2 is an NP-hard problem. It is important to note that with leakage power increasing as a fraction of the total power of a design, it is no longer sufficient to simply find the input vector that minimizes circuit leakage. It is arguably more important to find the leakage for *all* input vectors (of course, the minimum leakage vector can be found by this exercise). When comparing candidate implementations of a design with the same minimum leakage values, one would prefer the design that has a leakage histogram with the largest number of input vectors contributing lower leakage values. This would not only minimize the leakage during the regular operation of the circuit, but also ease the task of finding a vector that results in minimum leakage. It was reported in [9] that the maximum leakage value of a design can be as high as $2.4\times$ the minimum value ($1.6\times$ on average), again underscoring the importance of computing the leakage of *all* input vectors for implementations and choosing one with a favorable leakage histogram. Some of the existing work done in this area is discussed in the following section.

Table 3.1 Leakage of a NAND3 gate

Input	Leakage (A)
000	1.37389e−10
001	2.69965e−10
010	2.70326e−10
011	4.96216e−09
100	2.62308e−10
101	2.67509e−09
110	2.51066e−09
111	1.01162e−08

3.5 Related Previous Work

Several existing research works attempt to model and minimize the leakage currents in a combinational design. Some of these efforts [2,3,7–11,13,16,16] are described in Chap. 2.

All of the techniques cited above attempt to compute a *single* vector, which results in a minimum (or maximum) leakage state. An approach to compute the leakage values for *all possible input combinations* is presented in this chapter. Using ADDs [3,6], the leakage of the circuit for *all* input vectors is *implicitly* represented in a single structure. The inherent sharing of nodes in such a structure allows for a compact representation of the leakage of the design. In order to improve the efficiency of the leakage ADD construction, the values of the leaf nodes are binned so as to reduce the number of leaf nodes of the ADD. This reduces the number of discriminants[2] (as well as the number of nodes) in the leakage ADD of the design. The histogram of leakage values (constructed from the leakage ADD) is used for comparing candidate implementations of a circuit. In [5], the authors also present an ADD-based algorithm to determine the lowest leakage state of a circuit. They partition a circuit η into subcircuits and determine the minimum leakage value and the Minimum Leakage Vector (MLV) of each subcircuit. These leakage values are then summed in order to generate the minimum leakage value of η, and the MLV for η is generated by concatenation of the MLVs of the subcircuits. In the approach described in this chapter the entire range of leakage values are binned as opposed to pruning of all the leakage values except the minimum (or maximum) for the individual subcircuits. In [15], the authors use ADDs to find the leakage of a channel-connected region (CCR) as a function of its inputs. The focus in [15] was on full-custom circuitry and the authors used their technique to find functional failures in CCRs due to excessive leakage (input vectors that caused leakage to go above a certain value). Exclusivity constraints were added to constrain the ADD of a CCR to legal input vectors. We next describe the approaches for computing the exact and approximate leakage values for all input vectors for a circuit.

3.6 Our Approach

The approach described in this chapter is termed as AL^{all}. The exact version of AL^{all} is called AL^{all}_{ex}, while the approximate version is called AL^{all}_{app}.

3.6.1 Exact Computation of the Leakages of All Vectors

In order to compute the exact leakages of all vectors, the approach, called AL^{all}_{ex}, is described below. Consider a combinational logic network η, consisting of logic gates

[2] The number of discriminants of an ADD is the number of unique leaves of the ADD.

G_j selected from some library P. The ROBDD of G_j is referred to as g_j, and the leakage ADD of G_j as $\mathcal{G}_j$. This ADD represents the leakage value of each primary input minterm m of g_j (obtained by following the path from the root, indicated by the literals of m, until a terminal vertex is reached). The value of this vertex is the leakage of G_j under the input m. Note that the support of $\mathcal{G}_j$ is the primary inputs of the circuit.

Assume that for each gate G_j, there is an array called ($lkg\_array(G_j)$) describing its leakage values for all possible values of its immediate fanins. For example, if the G_j was a two-input gate, then its leakage array would consist of four values, corresponding to all four possible input combinations for the gate. Let the two fanins be called H_1 and H_2. For ease of the exposition, assume that these are sorted in a numerical order, so that the leakage value of the input combination 00 appears first, followed by that of the input values 01, and so on. Suppose that under some primary input minterm m, the ROBDDs h_1 and h_2 evaluate to $h1_{val}$ and $h2_{val}$ respectively. The corresponding leakage value for the gate G_j is found by indexing the $(h1_{val} : h2_{val})$th value of $lkg\_array(G_j)$. For example, if $h1_{val} = 1$ and $h2_{val} = 0$, the second value of $lkg\_array(G_j)$ is indexed to obtain the appropriate leakage value.

The algorithm AL_{ex}^{all} proceeds as follows. It first finds the ROBDDs of all network nodes. Next, it finds the (global) leakage ADDs of each of the nodes in the network using Algorithm 1. Suppose the leakage ADD of H is computed. Assume that it has two fanins F and G. The leakage ADD of H is found by the subroutine $node\_compute\_lkg\_ADD(f, g, lkg\_array(H))$. In this routine, if the ROBDDs f and g are constant (f_{val} and g_{val}, respectively), then the leakage value for this condition is simply found by indexing the $(f_{val} : g_{val})$th value of $lkg\_array(H)$ and returning an ADD node of this value. If either of f or g are non-constant, then the top variable v among these ROBDDs is returned. The computation recursively computes $\mathcal{H}_v$ and $\mathcal{H}_{\overline{v}}$, and finally returns $\mathcal{H} = ITE(v, \mathcal{H}_v, \mathcal{H}_{\overline{v}})$.

Algorithm 1 The node_compute_lkg_ADD algorithm

$node\_compute\_lkg\_ADD(f, g, lkg\_array(H)$
// terminal case below
if $f_{val} = is\_constant(f)$ && $g_{val} = is\_constant(f)$ **then**
  $\mathcal{H} = create\_ADD\_node(f_{val} : g_{val})$
  return $\mathcal{H}$
end if
$v = topvar(f, g)$
$f_v = cofactor(f, v)$
$f_{\overline{v}} = cofactor(f, \overline{v})$
$g_v = cofactor(g, v)$
$g_{\overline{v}} = cofactor(g, \overline{v})$
$\mathcal{H}_v = node\_compute\_lkg\_ADD(f_v, g_v, lkg\_array(H))$
$\mathcal{H}_{\overline{v}} = node\_compute\_lkg\_ADD(f_{\overline{v}}, g_{\overline{v}}, lkg\_array(H))$
$\mathcal{H} = ITE(v, \mathcal{H}_v, \mathcal{H}_{\overline{v}})$
return $\mathcal{H}$

Algorithm 1 is applicable for gates G_j with two inputs. The technology library usually consists of at most four-input gates. As a result, two additional routines similar to Algorithm 1 are required for three and four input gates.

Note that leakage ADDs of the mapped gates of the network need not be computed in any particular order. After the leakage ADDs of each gate have been computed, the leakage ADD of the entire circuit (this is referred to as $\mathcal{H}_{total}$), is found by adding each gate's leakage ADD. The routine to add two ADDs is shown in Algorithm 2. If the circuit has n gates, then this operation requires $n - 1$ ADD addition operations, since the addition of ADDs is performed in a pair-wise manner.

Algorithm 2 first tests if the ADDs $\mathcal{F}$ and $\mathcal{G}$ to be added are both constants. If this is the case (call the constants $\mathcal{F}_{val}$ and $\mathcal{G}_{val}$) it creates and returns an ADD node with value $\mathcal{F}_{val} + \mathcal{G}_{val}$. If at least one of $\mathcal{F}$ or $\mathcal{G}$ are non-constant, then the top variable v is found among them. $\mathcal{H}_v = add\_ADD(\mathcal{F}_v, \mathcal{G}_v)$ and $\mathcal{H}_{\overline{v}} = add\_ADD(\mathcal{F}_{\overline{v}}, \mathcal{G}_{\overline{v}})$ are recursively computed, and $\mathcal{H} = ITE(v, \mathcal{H}_v, \mathcal{H}_{\overline{v}})$ is returned.

Algorithm 2 The add_ADD algorithm

$add\_ADD(\mathcal{F}, \mathcal{G})$
// terminal case below
if $f_{val} = is\_constant(\mathcal{F})$ && $g_{val} = is\_constant(\mathcal{G})$ **then**
  $\mathcal{H} = create\_ADD\_node(\mathcal{F}_{val} + \mathcal{G}_{val})$
  return $\mathcal{H}$
end if
$v = topvar(\mathcal{F}, \mathcal{G})$
$\mathcal{F}_v = cofactor(\mathcal{F}, v)$
$\mathcal{F}_{\overline{v}} = cofactor(\mathcal{F}, \overline{v})$
$\mathcal{G}_v = cofactor(\mathcal{G}, v)$
$\mathcal{G}_{\overline{v}} = cofactor(\mathcal{G}, \overline{v})$
$\mathcal{H}_v = add\_ADD(\mathcal{F}_v, \mathcal{G}_v)$
$\mathcal{H}_{\overline{v}} = add\_ADD(\mathcal{F}_{\overline{v}}, \mathcal{G}_{\overline{v}})$
$\mathcal{H} = ITE(v, \mathcal{H}_v, \mathcal{H}_{\overline{v}})$
return $\mathcal{H}$

Once $\mathcal{H}_{total}$ (the sum of all the leakage ADDs of the gates in the design) is computed, the minimum valued leaf L_{min} (which is the minimum discriminant of $\mathcal{H}_{total}$) of the final ADD is found. This discriminant corresponds to the lowest leakage state of the design. A primary input vector that results in this leakage value is found by using Algorithm 3. A similar exercise can be conducted for any discriminant, which enables the construction of a leakage histogram for the design.

Algorithm 3 Finding an input vector with minimum leakage L_{min}

$find\_a\_minterm\_with\_min\_leakage(\mathcal{H}_{total})$
$\mathcal{H}_{thresholded} = ADD\_threshold(\mathcal{H}_{total}, L_{min} + \delta)$
$h_{thresholded} = ADD\_to\_BDD(\mathcal{H}_{thresholded})$
return $BDD\_find\_minterm(h_{thresholded})$

Thresholding an ADD consists of the task of converting it into an ADD with fewer discriminants. *ADD_threshold*($\mathcal{H}, val$) makes all discriminants with values greater than or equal to val point to the 0 discriminant. All discriminants with values less than val are retained in the result.

Algorithm 3 first thresholds $\mathcal{H}_{total}$ with the value $L_{min} + \delta$. The value δ is such that there is no leakage value for the design in the closed interval $[L_{min}, L_{min} + \delta]$. In other words, there is no discriminant in the leakage ADD $\mathcal{H}_{total}$ in the above closed interval. Therefore, the resulting leakage ADD after thresholding ($\mathcal{H}_{thresholded}$) consists of exactly two discriminants (L_{min} and 0). Next, $\mathcal{H}_{thresholded}$ is converted into a BDD, by replacing the L_{min} discriminant by the 1 discriminant. A path to the 1 terminal node in this BDD is now found by using the well-known linear-time BDD algorithm to find a single minterm.

In a similar manner, the BDD for any specific leakage value (i.e. any specific discriminant of the leakage ADD) can be found. For a general leakage value L other than the maximum or minimum, the thresholding with threshold values $L + \delta$ as well as $L - \delta$ needs to be done, where δ is such that there is no other discriminant of the leakage ADD in the interval $[L + \delta, L - \delta]$. From the resulting BDD of the result, the standard linear-time BDD algorithms can be used to find the number of minterms for the discriminant of value L. From this, the leakage histogram for the circuit is computed.

The CUDD [1] package is used for all the ADD operations in this chapter. This package has routines to perform the operations described in the algorithms described in this approach.

3.6.2 Approximate Computation of Leakages of All Vectors

The algorithm AL_{ex}^{all} of Sect. 3.6.1 produces the exact leakage values for the circuit being considered. Also, the BDD representation of all minterms with any specific leakage value L can be computed as described in Sect. 3.6.1. From this BDD, the number of input vectors (or a single vector) with leakage L can be computed in linear time. However, in an exact ADD representation of circuit leakage, the number of discriminants can be quite large. As a consequence, it is important to compute the circuit leakage ADDs in an *approximate* manner. This results in a reduction in the memory utilization and thereby allows the method to handle larger designs.

The algorithm AL_{app}^{all} computes the approximate leakage ADD of the circuit. In this approach the discriminant values are discretized during the *add_ADD* operation, such that the total number of discriminants of the added result are bounded by a user-specified constant m. The following subsection elaborates upon the discretization approach.

3.6.2.1 Binning of Leakage ADD Values

Since the library used consists of gates with up to four inputs, the maximum number of discriminants for the leakage ADDs of any gate is limited to 16. However, the resulting ADD after the *add_ADD* operation on two ADDs with D_1 and D_2 discriminants, respectively may have as many as $D_1 \cdot D_2$ discriminants. To control the size of the resulting ADD after addition, discretization of the discriminants of the result is performed. The discretization is driven by a user-specified constraint m, which represents the maximum number of discriminants in any ADD constructed (intermediate or final).

Consider the addition of two ADDs $\mathcal{F}$ and $\mathcal{G}$, using the *add_ADD* routine. Let the minimum and maximum discriminant values of $\mathcal{F}$ ($\mathcal{G}$) be $L^{\mathcal{F}}_{min}$ and $L^{\mathcal{F}}_{max}$ ($L^{\mathcal{G}}_{min}$ and $L^{\mathcal{G}}_{max}$), respectively. As a consequence, the minimum and maximum discriminant values of the result will be $(L^{\mathcal{F}}_{min} + L^{\mathcal{G}}_{min})$ and $(L^{\mathcal{F}}_{max} + L^{\mathcal{G}}_{max})$, respectively. Let the interval between these two values be R. Next discretize the interval into m values $(L^{\mathcal{F}}_{min} + L^{\mathcal{G}}_{min})$, $(L^{\mathcal{F}}_{min} + L^{\mathcal{G}}_{min} + \frac{R}{m-1})$, $(L^{\mathcal{F}}_{min} + L^{\mathcal{G}}_{min} + \frac{2R}{m-1})$, $(L^{\mathcal{F}}_{min} + L^{\mathcal{G}}_{min} + \frac{3R}{m-1})$, $\ldots$, $(L^{\mathcal{F}}_{min} + L^{\mathcal{G}}_{min} + \frac{(m-2)R}{m-1})$, $(L^{\mathcal{F}}_{max} + L^{\mathcal{G}}_{max})$.

Next, during the terminal case computation of Algorithm 2, compute $v = \mathcal{F}_{val} + \mathcal{G}_{val}$ and adjust its value to the nearest of the m discretized discriminant values described in the previous paragraph. Let the adjusted value be v_{adj}. Then, the value returned by Algorithm 2 in the terminal case is v_{adj}.

This limits the total number of discriminants in the result of *add_ADD* to m, instead of $D_1 \cdot D_2$, resulting in significantly reduced memory utilization in general. Also, the maximum error introduced by a single step of this addition is $\frac{1}{2(m-1)}$, allowing the user to trade off the memory utilization and maximum tolerable error.

3.6.2.2 Extensions to the Approach

In its current form, this algorithm computes the leakage ADDs for up to medium-sized circuits. To improve this further, a partitioned [12] construction of leakage ADDs may prove beneficial. In this approach, a k-way min-cut partitioning of the circuit is first performed, and the leakage ADDs of each partition are computed separately (on the space of the local inputs for that partition), before finally computing the image of these ADDs on the space of the primary inputs of the design.

Another application of this approach would be to compute the leakage ADD $\mathcal{G}$ for an arithmetic unit, from the leakage ADD $\mathcal{G}_s$ of a bit-slice of the unit. Suppose that the ith bit slice depends on free variables[3] v^i_f and bound variables[4] v^i_b. Let the leakage ADD of the ith bit slice be $\mathcal{G}^i_s(v^i_b, v^i_f)$[5], and the leakage ADD of the logic driving variables v^i_b be called g^i_b. The leakage ADD $\mathcal{G}$ can be computed by Algorithm 4.

In this manner, the total leakage of the arithmetic unit is computed iteratively, using the computed leakage ADD of a single slice. In the ith iteration, each bound variable is substituted in the leakage ADD of the ith slice with the leakage ADD

[3] Free variables are variables that are primary inputs of $\mathcal{G}$.

[4] Bound variables are variables of $\mathcal{G}_s$ that are the outputs of other bit slices in the design.

[5] $\mathcal{G}^i_s(v^i_b, v^i_f)$ is computed from the leakage ADD of a generic slice ($\mathcal{G}_s$) by a simple variable substitution.

Algorithm 4 Finding $\mathcal{G}$ from $\mathcal{G}_s^i$

$\mathcal{G} \leftarrow \mathcal{G}_s^1$
for $(i = 2; i <= n; i++)$ **do**
  $\mathcal{G} \leftarrow \mathcal{G} + BDD\_substitute(\mathcal{G}_s^i, v_b^i, g_b^i)$
end for
return $\mathcal{G}$

of the driving logic for that variable. The resulting leakage ADD of the slice is then added to the leakage ADD of the entire design. Hence, the computation of the leakage ADD of any slice i includes the constraints imposed by the leakage values for slices j whose outputs are inputs to the slice i.

3.7 Experimental Results

The technique AL_{app}^{all} was applied on a series of MCNC91 benchmark designs, using a 0.1-μm technology library with 13 gates, with between 1 and 4 inputs. After running technology-independent logic optimizations (*script_rugged* in SIS [14]), these designs were mapped for area and delay (again in SIS).

The AL_{ex}^{all} and AL_{app}^{all} leakage computation techniques were implemented in SIS and implemented using the CUDD [1] package. Applying the approximate technique AL_{app}^{all} with discretized discriminants enabled the computation of leakage ADDs for larger designs.

Tables 3.2 and 3.3 describe the maximum and minimum leakages (in pA) of four designs, as a function of the value of m (the number of discretized discriminants used during ADD construction). Each design was mapped for minimum area as well as minimum delay. The row labeled "exact" represents the leakages with no discretization of leakage values (effectively $m = \infty$). Note that a good choice of the values of m is between 12 and 16 for most cases.

Figure 3.6 describes the range of leakage values for the minterms mapped to the lowest discriminant of the ADD, compared against the normalized value of the range of the exact leakage. Ideally, this should be a point, with leakage L_{min}. It was observed that for most designs, this range is small, indicating that the method is accurate. The approximate experiments for this figure were performed with $m = 20$.

Table 3.4 reports the maximum and minimum leakage (represented in 10's of pA) for several designs, mapped both for minimum area as well as delay. It was observed that mapping for minimum area results on average in a 20% reduction in both the maximum and minimum leakage value, compared to delay mapping. The experiments in these tables were performed with $m = 12$.

The leakage histograms associated with the leakage ADDs were computed for some designs. For this experiment, $m = 20$ was used. The comparison between the area-mapped and delay-mapped histograms suggests that the area-mapped histograms are typically "better," with a larger number of minterms that have smaller leakage values. Figure 3.7 illustrates the results of this experiment.

Table 3.2 Accuracy vs. bin size I

	9symml				cc			
	Delay map		Area map		Delay map		Area map	
	min	max	min	max	min	max	min	max
exact	622.9	734.6	474.1	611.8	193.2	272.5	127.2	227
20 bins	540.8	772.3	429.1	633.2	209.6	267.8	131.5	221.1
16 bins	396.7	955.8	402.9	600.8	197.2	261.5	122	209.8
12 bins	285	1064.5	284	821.6	197.5	270.4	117.3	253.5
8 bins	212.4	1206.6	199.3	964.4	91	360.1	76.4	278
4 bins	212.4	1206.6	199.3	964.4	91	360.1	76.4	278

Table 3.3 Accuracy vs. bin size II

	decod				alu2			
	Delay map		Area map		Delay map		Area map	
	min	max	min	max	min	max	min	max
exact	187.8	238.6	30.6	79.9	1241.9	1382.9	872.8	1060.7
20 bins	200.8	239.1	31	83	905.5	1771.4	645.1	1348.5
16 bins	208	241.9	27.6	90.8	700.5	2005.2	576	1563.3
12 bins	212.6	235.5	23.6	74.9	536.7	2193.2	484.8	1753.4
8 bins	89.3	314.5	33	92.3	511.9	2251.2	382	1856.5
4 bins	89.3	314.5	33	92.3	511.9	2251.2	382	1856.5

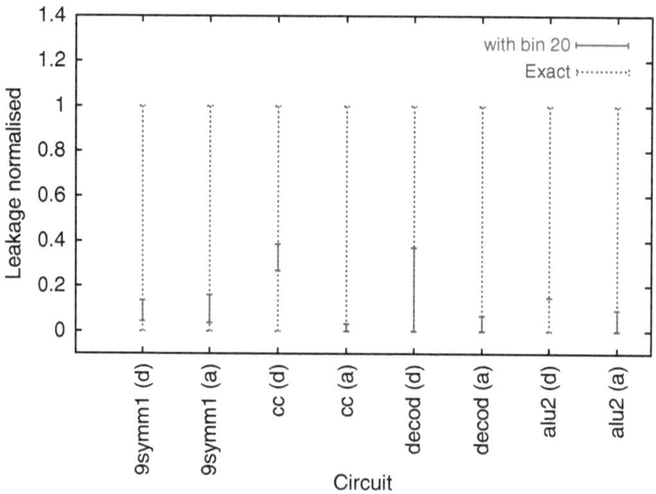

Fig. 3.6 Error of ADD-based leakage computation

Table 3.4 Leakage min/max values for area and delay-mapped designs

	Delay mapped		Area mapped	
	max	min	max	min
9symml	10645.4	2850.7	8216.3	2840.0
b9	6385.3	1385.7	5573.1	1333.5
c8	6542.8	2564.5	6572.5	2337.5
cc	2704.7	1975.5	2535.9	1173.6
cht	9589.2	3248.4	9077.8	3100.9
cm138a	1179.0	885.5	618.4	291.6
cm150a	2332.8	1078.2	2109.1	956.3
cm151a	1153.0	653.5	1153.0	653.5
cm152a	974.6	613.2	974.6	613.2
cm162a	2167.7	1213.4	2131.3	958.1
cm163a	1976.9	1189.4	2218.4	1067.8
cm42a	993.4	777.0	672.0	417.9
cm82a	1062.0	855.8	929.8	712.6
cm85a	2147.4	1245.9	1658.8	1084.7
count	7740.8	2354.9	6427.2	892.0
cu	2316.0	1328.5	1912.7	1091.8
f51m	3331.7	2562.7	3224.4	2255.6
frg1	7814.1	1723.1	7515.6	1298.4
i1	2453.0	785.8	1950.9	558.3
lal	5406.9	1400.9	4584.6	1004.8
majority	429.8	269.1	350.7	192.1
mux	2541.8	1672.0	2064.6	1088.3
parity	3031.0	1884.9	3031.0	1884.9
pcle	3982.1	1453.9	3578.4	1397.3
pcler8	5485.5	1527.7	4849.5	1352.9
pm1	2043.5	856.1	1763.3	504.1
sct	3730.8	1729.9	3136.4	1618.2
t	321.8	179.7	321.8	179.7
tcon	1465.8	1052.5	1070.2	656.9
unreg	5199.0	2893.4	5083.3	1966.5
x2	1557.5	704.9	1340.0	587.2
z4ml	1715.6	1389.2	1482.5	1051.8
decod	2355.1	2126.8	749.7	236.9
alu2	21932.5	5367.9	17534.8	4848.5
alu4	43888.3	10457.2	33218.0	7870.9
t481	48647.5	9664.6	38554.8	5936.5
vda	34696.6	11041.7	25198.8	7223.9
apex7	14949.1	3320.9	12413.3	1802.8
AVERAGE	7286.6	2323.3	5942.1	1711.7

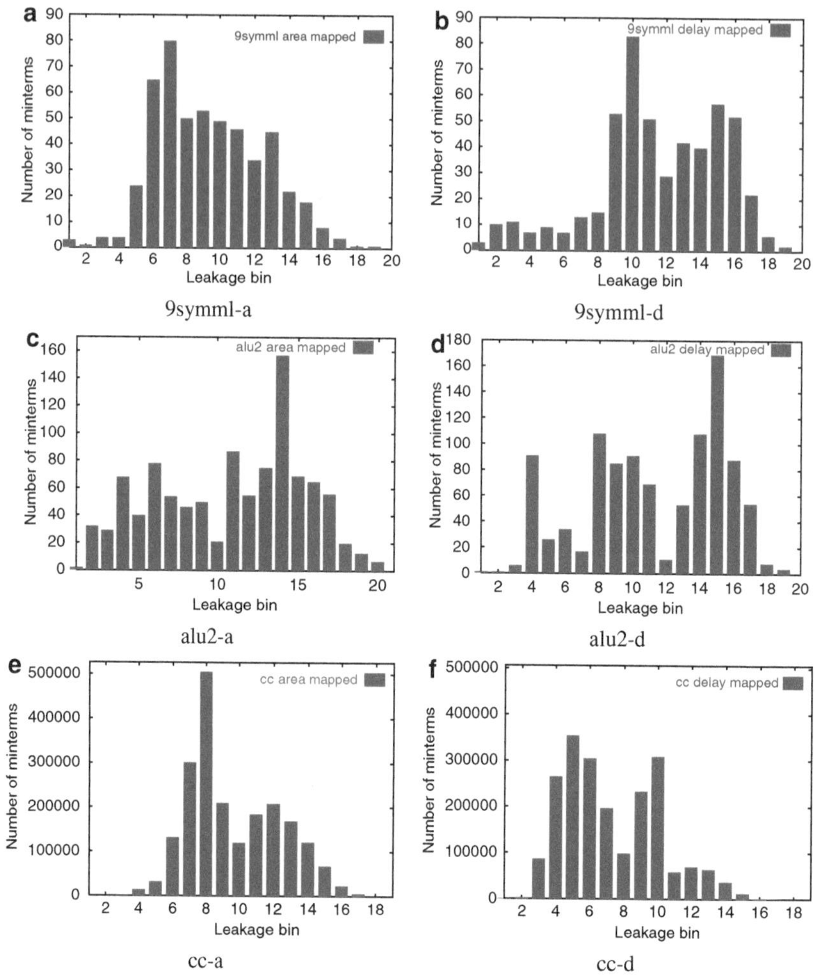

Fig. 3.7 Leakage histograms for delay and area-mapped circuits

3.8 Summary

This chapter described the algorithms used for computing the exact and approximate leakage values for all input vectors for a circuit. The intuition behind these algorithms was explained, along with an exposition of the details. In addition, some extensions for future work were discussed. The pseudo-code was provided for a peruse explanation of the algorithms. Further, results obtained for the approximate leakage ADD (computed with varying number of discriminants) are compared with exact values. In addition, two different implementations, mapped for *area* and *delay*, for some design are compared. The comparison is made on the different leakage histograms obtained for the above two common mapping criteria.

References

1. CUDD: CU Decision Diagram Package. http://vlsi.colorado.edu/ fabio/CUDD/cuddIntro.html
2. Aloul, F., Hassoun, S., Sakallah, K., Blauuw, D.: Robust SAT-Based Search Algorithm for Leakage Power Reduction. In: Proc. Power and Timing Models and Simulation. Seville, Spain (2002)
3. Bahar, R.I., Frohm, E.A., Gaona, C.M., Hachtel, G.D., Macii, E., Pardo, A., Somenzi, F.: Algebraic Decision Diagrams and Their Applications. Formal Methods in Systems Design **10**(2/3), 171–206 (1997)
4. Cao, Y., Sato, T., Sylvester, D., Orshansky, M., Hu, C.: New Paradigm of Predictive MOSFET and Interconnect Modeling for Early Circuit Design. In: Proc. IEEE Custom Integrated Circuit Conference, pp. 201–204. Orlando, FL (2000). http://www-device.eecs.berkeley.edu/ ptm
5. Chopra, K., Vrudhula, S.: Implicit Pseudo Boolean Enumeration Algorithms for Input Vector Control. In: Proc. Design Automation Conference, pp. 767–772. San Diego, CA (2004)
6. Clarke, E.M., McMillan, K.L., Zhao, X., Fujita, M., Yang, J.: Spectral transforms for large boolean functions with applications to technology mapping. In: Proceedings of the 30th International Conference on Design Automation, pp. 54–60. ACM Press (1993). DOI http://doi.acm.org/10.1145/157485.164569
7. Duarte, D., Tsai, Y., Vijaykrishnan, N., Irwin, M.J.: Evaluating Run-Time Techniques for Leakage Power Reduction. In: 7th ASPDAC/15th International Conference on VLSI Design (2002)
8. Ferre, A., Figueras, J.: Characterization of Leakage Power in CMOS Technologies. In: Proc., IEEE International Conference on Electronics Circuits and Systems, pp. 85–188 (1998)
9. Gao, F., Hayes, J.: Exact and Heuristic Approaches to Input Vector Control for Leakage Power Reduction. In: Proc. International Conference on Computer-Aided Design, pp. 527–532. San Jose, CA (2004)
10. Halter, J., Najm, F.: A Gate-Level Leakage Power Reduction Method for Ultra Low Power CMOS Circuits. In: Proc. Custom Integrated Circuits Conference, pp. 475–478. Santa Clara, CA (1997)
11. Johnson, M., Somasekhar, D., Roy, K.: Models and Algorithms for Bounds on Leakage in CMOS Circuits. IEEE Transactions on Computer-Aided Design of Integrated Circuits and Systems **18**(6), 714–725 (1999)
12. Narayan, A., Jain, J., Fujita, M., Sangiovanni-Vincetelli, A.: Partitioned ROBDDs–A Compact, Canonical and Efficiently Manipulable Representation for Boolean Functions. In: Proceedings, IEEE/ACM International Conference on Computer-Aided Design, pp. 547–554 (1996)
13. Rao, R., Liu, F., Burns, J., Brown, R.: A Heuristic to Determine Low Leakage Sleep State Vectors for CMOS Combinational Circuits. In: Proc. International Conference on Computer-aided Design, pp. 689–692. San Jose, CA (2003)
14. Sentovich, E.M., Singh, K.J., Lavagno, L., Moon, C., Murgai, R., Saldanha, A., Savoj, H., Stephan, P.R., Brayton, R.K., Sangiovanni-Vincentelli, A.L.: SIS: A System for Sequential Circuit Synthesis. Tech. Rep. UCB/ERL M92/41, Electronics Research Laboratory, Univ. of California, Berkeley, CA 94720 (1992)
15. Song, H.Y., Bohidar, S., Bahar, R.I., Grodstein, J.: Symbolic Failure Analysis of Custom Circuits due to Excessive Leakage Current. In: Proc. IEEE International Conference on Computer Design, pp. 70–75 (2003)
16. Zhanping, C., Johnson, M., Liqiong, W., Roy, W.: Estimation of Standby Leakage Power in CMOS Circuit Considering Accurate Modeling of Transistor Stacks. In: Proc. International Symposium on Low Power Electronics and Design, pp. 239–244. Monterey, CA (1998)

Chapter 4
Finding a Minimal Leakage Vector in the Presence of Random PVT Variations Using Signal Probabilities

4.1 Overview

The control of leakage power consumption is a growing design challenge for current and future CMOS circuits. A heuristic approach (referred to as MLVC) is to determine the input vector that minimizes leakage for a combinational design. This approach utilizes approximate signal probabilities of internal nodes to aid in finding a minimal leakage vector. We utilize a probabilistic heuristic to select the next gate to be processed as well as to select the best state of the selected gate. A fast Boolean Satisfiability (SAT) solver is employed to ensure the consistency of the assignments that are made in this process. A variant of MLVC, referred to as MLVC-VAR, is also presented. MLVC-VAR includes the effect of random variations in leakage values due to process, voltage and temperature (PVT) variations. Including the effect of PVT variations for determining minimum leakage vector is crucial because leakage currents have an exponential dependence on power supply, threshold voltage and temperature. Experimental results indicate that our MLVC method has very low runtimes, with excellent accuracy compared to existing approaches. Further, the comparison of the mean and standard deviation of the circuit leakage values for MLVC with MLVC-VAR and an existing random vector generating approach proves the need for considering these variations while determining the minimum leakage vector. MLVC-VAR reports, on average, about 9.69% improvement over MLVC with similar runtimes and 5.98% improvement over the random vector generation approach with significantly lower runtimes.

The remainder of this chapter is organized as follows: The motivation for this work is described in Sect. 4.3. Section 4.4 discusses some previous work in this area. In Sect. 4.5 we describe our signal probabilities-based approach, and discuss its experimental results in Sect. 4.6. Conclusions and future work are discussed in Sect. 4.7.

N. Jayakumar et al., *Minimizing and Exploiting Leakage in VLSI Design*,
DOI 10.1007/978-1-4419-0950-3_4, © Springer Science+Business Media, LLC 2010

4.2 Introduction

An efficient heuristic to determine the minimum leakage vector (i.e. the input vector that drives the circuit to its lowest leakage state) is proposed in this chapter. This problem can be viewed as one of selecting the state of each gate in the circuit such that the total leakage over all gates is minimized, and the state of each gate in the circuit is logically feasible (i.e. is logically compatible with states of all the other gates). In this chapter, we present a heuristic approach (referred to as MLVC) to determine the input vector that minimizes leakage for a combinational design. The distinguishing feature of our approach is that it is guided by signal probabilities. In other words, the selection of the best candidate gate, as well as the input state to use for that gate, is performed probabilistically. The intuition behind such selections is that they have a high likelihood of resulting in a circuit state that is logically justifiable, while minimizing leakage as well. Additionally, the effect of PVT variations can be elegantly incorporated into such a probabilistic formulation.

With the decrease in process feature sizes, the effect of PVT variations has become significant. Since sub-threshold leakage has a critical dependency on temperature, power supply, channel length and threshold voltage, the PVT variations heavily influence the leakage values and correspondingly the minimum leakage vector determination. In [23], the authors experimentally prove that a simple assumption of uniform temperature and power supply variation can underestimate the full chip leakage by 30%. In [18], the authors establish the importance of considering the variations of within-die threshold voltage and channel length for accurate sub-threshold leakage current prediction. They determine that the sub-threshold leakage power can be underestimated or overestimated by 1.5× to 6.5× by ignoring these within-die variations.

Keeping the significant dependence of leakage on PVT variations in mind, a variant of MLVC, called MLVC-VAR, is also presented. *MLVC-VAR includes the effect of random PVT variations while determining the Minimum Leakage Vector (MLV).* Currently our approach does not account for correlation between the PVT variables. However, these correlations can be easily incorporated into the approach as we describe in the sequel. The effect of PVT variations is considered in the formulation of both the heuristics: for selecting the best candidate gate and the best leakage state for that gate. To the best of our knowledge, no other work on MLV determination to date has considered these important variations in its formulation.

In our experiments, we compare the accuracy and runtimes for MLVC with other existing techniques for determining the MLV. Further, we compare the mean circuit leakage and the standard deviation of the circuit leakage of the input vectors determined by MLVC and MLVC-VAR.

4.3 The Intuition Behind Our Approach

As mentioned previously in Chap. 2, input vector control is an effective technique with little or no circuit modification to reduce leakage currents in a combinational design. Further, including the effect of PVT variations for determining minimum leakage vector is crucial because leakage currents have an exponential dependence on power supply, threshold voltage and temperature. To the best of the authors' knowledge, no other minimum leakage vector determination work has to date included the effect of PVT variations. This chapter addresses intra-die variations. Intra-die variations are an important contributor to the mean leakage and the standard deviation of leakage for a combinational circuit. Since the intra-die variations of a gate are dependent on the logic state of the gate [3,21], we propose the following objective for the MLVC-VAR approach: We aim at reducing the mean leakage plus six times the standard deviation (cost function = $\mu + 6\sigma$) of the combinational circuit, by choosing the input vector that sets the logic states of all the gates in the most "favorable" manner (conducive to lowering the cost function). It can be conjectured that considering intra-die variations just leads to an increased expectation value, but the best state remains the best state. By this reasoning, optimization without intra-die PVT will lead to nearly the same result. The following example explains why this conjecture is false:

The mean, nominal and standard deviation of the logic gates (Inverter, NOR2 and AND2) for different logic states are listed in Table 4.1. Note that the leakage values are from the library we used in our experiments.

Consider the circuit in Fig. 4.1 that is composed of three logic gates. The output d of the circuit evaluates to $\overline{a} + b \cdot c$. In case of MLVC (or an MLV determination technique that only aims at reducing the nominal circuit leakage) the best input

Table 4.1 Mean, nominal and standard deviation for the logic gates

Inverter				
Input	Output	μ (nA)	Nominal (nA)	σ (nA)
0	1	1.8832	2.2904	1.5055
1	0	3.7881	6.5253	8.2548
NOR2				
Input	Output	μ (nA)	Nominal (nA)	σ (nA)
00	1	3.7668	4.5818	3.0284
01	0	4.4738	7.0279	7.7904
10	0	7.5724	13.0926	16.7359
11	0	0.4468	0.5574	0.3854
AND2				
Input	Output	μ (nA)	Nominal (nA)	σ (nA)
00	0	3.9742	6.7527	12.5603
01	0	7.0834	10.5649	11.3243
10	0	5.5780	8.6271	7.5410
11	1	9.4602	15.4030	10.3201

Fig. 4.1 Example circuit for motivating MLVC-VAR

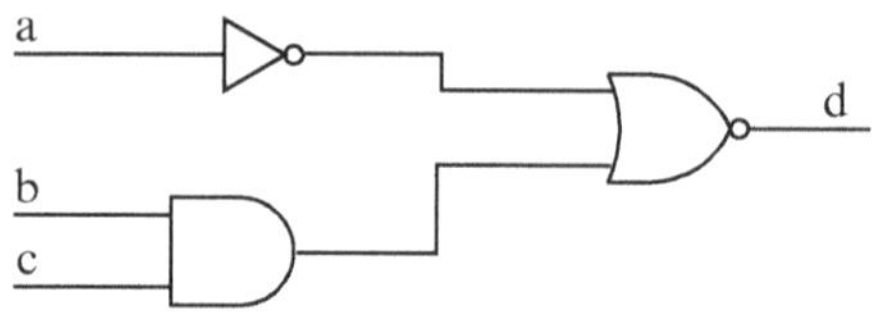

vector would be 000 (i.e. the assignment of 0 to all three inputs a, b and c). From Table 4.1, in this case the total nominal leakage of the circuit would be 16.0710 nA. Similarly, the metric $\mu + 6\sigma$ in this case would be 141.4684 nA.

On the other hand, MLVC-VAR aims at reducing the metric $\mu + 6\sigma$ as opposed to only reducing the nominal leakage for the combinational circuit. In this case the best input vector assignment would be a $=$ 0, b $=$ 1 and c $=$ 1. Again the values are computed using Table 4.1. In this case even though the nominal leakage of the circuit would be 18.2508 nA (which is 11.94% higher than that reported from MLVC) the metric $\mu + 6\sigma$ in this case would be 85.0562 nA (which is 66.32% lower than that reported using MLVC).

This example explains why MLVC in the presence of intra-die variations would not be adequate. MLVC alone might possibly yield a vector for which the worst case ($\mu + 6\sigma$) leakage of the combinational circuit is higher than what MLVC-VAR would compute.

4.4 Related Previous Work

Some of the existing works that address input vector control ([1,4,10,13,14,19,26]) are described in Chap. 2. In contrast to the approaches of [10, 13, 26], the approach described in this chapter is a heuristic that uses signal probabilities and leakage values of the gates to help assign values to the nodes in a combinational circuit.

Similar to the approaches described in [1, 4], the approach described for MLV determination in this chapter requires a SAT solver as well, but does not involve internal node modifications, which makes it computationally tractable. Moreover, larger designs are handled more easily, since the SAT solver is invoked only to verify the state assignments of individual gates, after every k iterations. The frequency with which the SAT solver is invoked is decided using experimental data, in order to run large circuits with low run-times and good accuracy. In [19], the authors present a greedy search-based heuristic, guided by node controllabilities, controllability lists and functional dependencies. In our approach, we do not compute node controllabilities or their controllability lists. We compute signal probabilities instead, which are computed in time that is linear in circuit size.

In [24], the authors describe several methods to set pass/fail limits for IDDQ testing, among which is a probabilistic method. For each cell in a design (each cell is assumed to have a single output, implemented in static CMOS), the authors compute the maximum IDDQ when the output is ON (OFF), assuming 4σ process variation limits. Additionally, the cell probabilities are determined for the input vectors that result in the maximum IDDQ of the cell for both the ON and OFF states. In contrast to [24], the approach in this chapter takes into account probabilities of all input vectors of a cell implicitly, and not just those of two outputs that result in a worst-case IDDQ value. Further, the signal probabilities, in the heuristic presented in this chapter, are adjusted for reconvergence, unlike those presented in [24].

Once the minimum leakage input pattern is found, this vector is used to drive the circuit in its standby mode. This may require the addition of a number of multiplexers at the primary inputs of the circuit. The multiplexers are controlled using a sleep signal (in a scan-based design, these multiplexers are not required). Since the power reduction using these techniques can be achieved only for sleep durations that are sufficiently long, the sleep signal is activated only if the sleep duration is long enough.

All of the above cited MLV determination approaches ignore the within-die PVT variations. Some works that estimate leakage values considering these variations are discussed below.

The authors in [23] establish (by using iterative numerical methods) the dependency of leakage current on temperature and power supply and prove that an assumption of uniform temperature and power supply variation can underestimate the full chip leakage by 30%. In [18], the authors find that the sub-threshold leakage power can be overestimated or underestimated by 1.5×–6.5× if variations of within-die threshold voltage and channel length values are ignored. The authors of [25] present a probabilistic framework for full-chip estimation of leakage power distribution considering inter and intra-die process and temperature variations. A method for analyzing leakage current under process parameter variations including spatial correlations can be found in [7]. On the other hand, in [20], the authors develop an analytical expression to estimate the PDF of the leakage current and thence, estimate the variation in leakage current due to gate length and process variability.

The authors of [15] propose a projection-based algorithm to estimate the full-chip leakage power, considering both inter-die and intra-die variations, by extracting a low-rank quadratic model. In [5], the authors present a gate-sizing methodology to minimize the leakage power in the presence of process variations. They formulate a geometric programming problem by modeling leakage as a posynomial function.

Most of these papers provide a mathematical or probabilistic framework for estimation of circuit leakage current in the presence of PVT variations, and hence the leakage power consumption. Our heuristic, MLVC-VAR, in contrast, considers the PVT variations *while determining the MLV*. No existing work on MLV determination considers PVT variations. An extended abstract of the MLVC work presented in this paper can be found in [12]. It contains the details of MLVC but does not discuss MLVC-VAR and its results. Further, the runtimes reported for MLVC in [12] have been improved by 10× by careful modifications in the algorithm, which is discussed in the sequel.

4.5 Our Approach

The outlines of MLVC and MLVC-VAR are as follows:

- First, for both MLVC and MLVC-VAR, we compute signal probabilities for all nodes in the design, assuming that all inputs have a signal probability of 0.5. These probabilities are heuristically adjusted for inaccuracies arising from reconvergent fanouts.
- Next, we select the best candidate gate whose leakage we would like to set in a given iteration. For both MLVC and MLVC-VAR, this is performed by selecting the gate that is probabilistically most likely to result in the largest leakage reduction. For MLVC-VAR, we consider (in addition to the probabilistic signal values) the mean and standard deviation of the leakages at each state,[1] of each gate, before choosing a gate that results in the lowering of the standard deviation of the circuit leakage.
- We next select the best state for the chosen gate. In MLVC, for the gate thus selected we next assign its best state such that the leakage of the selected gate is probabilistically minimized. In MLVC-VAR, this state is chosen by considering not only the signal probabilities and leakage values, but also the standard deviation of the leakage values due to PVT variations. All other gates in the circuit, which are newly implied by the state just selected, are accounted for while making this decision. Prior to computing the cost metric for this step, we first test if the candidate state is consistent with the assignments made in the previous runs.
- In both MLVC and MLVC-VAR, we next test if the logic values that were set to 1 or 0 during this iteration are satisfiable, by calling a Boolean Satisfiability (SAT) solver. The SAT [9] problem can be defined as follows. Given a set V of variables, and a collection C of Conjunctive Normal Form (CNF) clauses over V, the SAT problem consists of determining if there is a satisfying truth assignment for C. For any circuit, one can potentially generate a CNF to represent the circuit [22]. In our method, the SAT solver is called on the CNF of the circuit to test if the currently assigned logic values are consistent with the circuit. Further, the SAT solver is called every p iterations to reduce the runtime. If the circuit is unsatisfiable, we undo the assignments of the last p iterations and find the iteration that caused the circuit to become unsatisfied. After making a different selection for that iteration, we proceed as before.
- After any iteration, for both MLVC and MLVC-VAR, gate probabilities are adjusted to account for the nodes that were newly assigned fixed logic values.
- A fixed number of passes are made for the circuit, with the above steps being applied successively. Each pass is more "lenient" in setting a node to a logic value v when its signal probability is different from those of v. The last pass is most lenient, allowing any v to be accepted. This feature is common to both MLVC and MLVC-VAR.

---

[1] A state of a gate stands for an assignment of the logical (1 or 0) value at each of its input, and hence a logical value assigned at its output. For example, a three-input gate will have eight states.

Algorithm 5 describes the pseudo-code for MLVC, for a combinational network η. The algorithm for MLVC-VAR is identical to Algorithm 5, except for the functions *find_best_gate*(η) and *find_best_leakage_state*(G, η). The differences are detailed in the following subsections.

Algorithm 5 Pseudo-code of MLVC

```
compute_minimum_leakage_vector(η, p){
compute_signal_probabilities(η)
finalvalues ← Φ
for i = 1; i ≤ k; i + + do
   temporaryvalues ← Φ
   iteration = 1
   G = find_best_gate(η)
   if (G is not marked visited) then
      S = find_best_leakage_state(G, η)
      if S satisfies m_i then
         temporaryvalues ← temporaryvalues ∪ S ∪ get_implications(S)
         propagate probabilities in TFO of temporaryvalues nodes
      end if
      if iteration is a multiple of p OR all inputs assigned/implied then
         if temporaryvalues are satisfiable then
            if all inputs assigned then
               exit
            end if
            finalvalues ← finalvalues ∪ temporaryvalues
         else
            temporaryvalues ← finalvalues
         end if
      end if
   end if
   iteration += 1
end for
}
```

4.5.1 Computing Signal Probabilities

The algorithm *compute_minimum_leakage_vector*(η) for both MLVC and MLVC-VAR begins by computing signal probabilities for all nodes in the network η.

Definition 4.1. Signal probability of a node X is the probability of X being at logic level "1".

The inputs are assumed to have probabilities of 0.5, and these probabilities are propagated throughout the circuit. If the input i of an n-input AND gate has probability p_i, then the output has probability $\Pi_i\, p_i$. Likewise, for an OR gate, the output has probability $1 - \Pi_i(1 - p_i)$. The probabilities of other gates can be found in

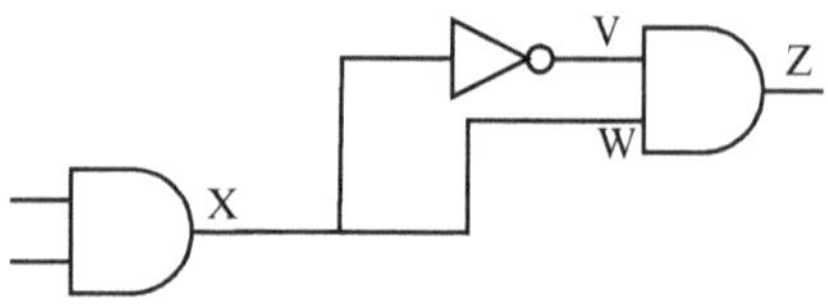

Fig. 4.2 Adjusting probabilities for reconverging nodes

a similar fashion. After the initial pass of propagation, we heuristically adjust for reconvergent fanouts. The heuristic for probability adjustment in the presence of reconvergence is explained with the help of Fig. 4.2.

Suppose a node X, with a statically computed probability of P_X reconverges at Z. Then we set the signal probability of X to 1 and 0, and find the probabilities of the inputs to the reconvergent gate (V and W). Suppose the probabilities of V (W) are V_1 (W_1) and V_0 (W_0), respectively, when X is set to 1(0). In this case, the new probability of Z is $P_Z^{new} = \frac{V_0 \cdot W_0 + V_1 \cdot W_1}{2}$.

From this we compute the adjustment factor for the probability of Z, as follows. Note that the adjustment factor is computed exactly once, in the beginning of the procedure.

$$Adjustment(Z) = \frac{(P_Z^{new} - P_Z)}{P_Z}$$

The physical meaning of P_Z^{new} is explained as follows. Assuming that 0 and 1 are equally likely at X, we can say that the signal probability of Z is $P_Z^{new} = \frac{V_0 \cdot W_0 + V_1 \cdot W_1}{2}$. If there was no reconvergence, then P_Z^{new} would be identical to P_Z. In the presence of reconvergence, however, P_Z^{new} deviates from P_Z by an amount equal to $Adjustment(Z)$.

In future updates of the probability of the node Z, suppose the statically computed probability of node Z is $P_Z^{modified}$. In that case, the final adjusted value of the probability of node Z is

$$P_Z^{adj} = (P_Z^{modified}) \cdot (1 + Adjustment(Z)).$$

In other words, $Adjustment(Z)$ is computed once and utilized to adjust the statically computed values of the probability of node Z, each time it is modified due to other assignments in the circuit.

In the example of Fig. 4.2, $Adjustment(Z) = -1$. Therefore, $P_Z^{adj} = 0$ each time the probability of Z is modified. This is reasonable, given that the output Z is logically 0.

If an adjustment of the probability of a node results in its probability becoming higher than $P_{threshold}$ (lower than $1 - P_{threshold}$), then the probability of the node is capped at $P_{threshold}$ ($1 - P_{threshold}$), respectively.

4.5.2 Finding the Best Leakage Candidate

Once signal probabilities are computed, we next select the best candidate gate whose input state we would like to finalize. For MLVC, gates are ranked by the probabilistic criterion:

$$C = \frac{\sum(p_i \cdot l_i)}{\sum(p_i)}(l_i^{\max} - l_i^{\min}).$$

Here, p_i is the probability that the gate is in state i. By "state," we mean a complete assignment of the inputs of the gate. The quantity l_i is the nominal leakage of the state i. The value $l_i^{\max}$ ($l_i^{\min}$) is the maximum (minimum) nominal leakage value of this gate. The gate with the maximum value of C is selected. In other words, this criterion selects gates that have a high probability of being in a high-leakage state. The last term in the expression for C ensures that gates with large leakage ranges are favored, since they offer potentially greater optimization flexibility. The gate that maximizes C is selected preferentially over others.

Note that due to the "snapping" of any signal probability higher (lower) than $P_{\text{threshold}}$ $(1 - P_{\text{threshold}})$ to $P_{\text{threshold}}$ $(1 - P_{\text{threshold}})$, no node can have signal probabilities identically equal to 1 or 0. Hence, there are instances when the sum of probabilities of all states of a node does not sum to unity and therefore the denominator in the above expression is not replaced by unity.

For MLVC-VAR, the expression is further biased to select a gate for which the standard deviation in the leakage values due to PVT variations is maximum. This biasing favors the selection of a gate that has higher variations in the leakage values. This is reasonable because in the next step, this gate is set to a state (among the possible states) that minimizes leakage as well as the leakage variations. Hence, it helps in avoiding a large standard deviation in the expected overall circuit leakage. Therefore, the final expression for C_{var} for MLVC-VAR is as follows:

$$C_{\text{var}} = \frac{\sum(p_i \cdot l_i \cdot r_i)}{\sum(p_i)}(l_i^{\max} - l_i^{\min}).$$

Here r_i is the range of the leakage value of a gate in state i. This range accounts for the PVT variations. Note that the l_i in this expression is the *mean* leakage of the state i, unlike for MLVC, where l_i is the nominal leakage of state i. Again, the gate that maximizes C_{var} is selected preferentially over others.

4.5.3 Finding Best Leakage State for Selected Gate

Suppose a gate G was selected by the previous step. For MLVC, we now want to assign it a state such that its leakage is minimized. This is done by applying the probabilistic criterion L below. Note that all gates other than G whose states become fully assigned[2] on account of implying the current state of G are also included in the computation of L. Let the number of such states be n. The value of probabilistic leakage in the numerator of L is normalized with respect to the number of such states and is computed as follows:

$$L = \frac{\sum_j (d_j \cdot l_j)}{n}.$$

L is computed for only those states whose assignments are consistent with the assignments made in the earlier passes. This test is done by invoking the BerkMin [11] satisfiability solver. If the assignment of a state s fails the test, we proceed to the next state for G, else (i.e. s is "legal") we compute L for s. Among all the legal states of the gate G, the state that minimizes L is preferentially selected over others. Here d_j is the *distance* of the values assigned to the gate inputs from their probabilistic values. For example, consider an AND gate with inputs a and b with probabilities 0.1 and 0.7, respectively. If inputs a and b in state j are logic 1 and logic 0, respectively, then the distance d_j is $(|1 - 0.1|)(|0.7 - 0|)$. l_j is the nominal leakage of state j. Note that the probability of a or b can never be exactly 1 or 0, because probability values higher (lower) than $P_{\text{threshold}}$ $(1 - P_{\text{threshold}})$ are snapped to $P_{\text{threshold}}$ (or 1-$P_{\text{threshold}}$). Hence d_j can never be exactly 0.

By minimizing L, we choose a state that has the lowest distance from its current probabilities, and because these probabilities are updated to account for the logic and for the structure of the circuit, this state would reduce the chances of assigning logically conflicting states. In order to bias the state selection towards assignments with lower leakage the distance is incremented by a value β. Likewise, in order to bias the state selection towards those with lower distance, we increment l_j by a fixed value γ. The relative values of β and γ are selected based on the relative scale of d_j and l_j values. In practice these values are determined experimentally.

Therefore, the modified value of L that is used is

$$L = \frac{\sum_j (d_j + \beta) \cdot (l_j + \gamma)}{n}.$$

For MLVC-VAR, analogous to L above, we utilize the selection criterion L_{var}. It is computed as follows:

$$L_{\text{var}} = \frac{\sum_j (d_j + \beta) \cdot (l_j + \gamma) \cdot (r_j + \rho)}{n}.$$

All variables in this expression denote the same values as the ones explained earlier in this sub-section, except that l_j is the mean (instead of nominal) leakage at state j. r_j is the range due to variations in the leakage values due to PVT variations. If the leakage distribution of a gate is $N(\mu_g, \sigma_g)$, then $r_j = 6\sigma_g$. Here, by

[2] A gate is said to be *fully assigned* if all its inputs are assigned to specific logic values.

minimizing L_{var}, we choose a state that in addition to minimizing leakage and the chances of a conflict minimizes the leakage range for that gate as well. The idea for opting for such a state is to reduce the circuit leakage and also minimize any variations in the expected overall circuit leakage.

Similar to the above biasing approach, in order to bias our selection towards assignments with lower leakage and lower distance, we increment the range by a fixed value ρ, in the computation of L_{var}.

4.5.4 Accepting Leakage States and Final MLV Determination

The state selected from the previous step is now implied throughout the transitive fanout (TFO) of the chosen gate. The resulting values are referred to as *temporary* values. The distance of the resulting implications is now checked against a margin value m_i. If any distance is greater than m_i, then the assignment to gate G is discarded. Initially, m_i is set to a small value, and with increasing iteration i, it is relaxed. This is in an attempt to get closer to a global minima, by a more careful selection of states in early iterations. We perform $k = 3$ iterations in our experiments.

Once the new implications are computed, the implied nodes' probabilities are adjusted to reflect the freshly computed implications. If a node is set to a logic 1, then its probability is set to $(1 - \alpha)$, while a node which is set to logic 0 has its probability updated to α.

For every p iterations (or if all primary inputs have been assigned or implied), we test if the *temporary* values are satisfiable (this test is done by invoking the BerkMin [11] satisfiability solver). If so, then all *temporary* values are designated as new *final* values, never to be modified in the future. If the *temporary* values are satisfiable, and all inputs are assigned, then the algorithm exits. If the *temporary* values could not be satisfied, then we roll back the *temporary* values, by copying the last set of *final* values into the set of *temporary* values. For up to the next p iterations, we call the satisfiability solver after each new state assignment. This is in an attempt to locate which of the last p assignments caused the unsatisfiability condition to occur. Once this state is identified, we again revert to calling the satisfiability solver after every p state assignments. If the satisfiability solver returns an unsatisfiable condition for a certain state s assigned at a particular gate g, then we never try assigning s to g again. An example explaining the invocation to a Boolean satisfiability solver is explained next.

Invoking a Boolean Satisfiability Solver: A combinational circuit can be represented in a Conjunctive Normal Form (CNF), which is the input format for most SAT solvers including BerkMin [11]. In our work, we invoke the SAT solvers in every few iteration to check the compatibility of intermediate assignments. This is done by augmenting the existing CNF for the combinational circuit with the clauses that represent the intermediate assignments.

For instance, a two-input AND gate with inputs A and B and output C such that C = A.B, in CNF consists of the following three clauses:

$$(C + A^{'} + B')$$
$$(C^{'} + A)$$
$$(C^{'} + B)$$

Suppose our intermediate assignments on the different variables are as follows:

$$\mathrm{A} = 1, \mathrm{B} = 0 \text{ and } \mathrm{C} = 0.$$

Now, to check the consistency of these assignments with the circuit we add the following clauses to the original CNF formula:

$$(A)$$
$$(B^{'})$$
$$(C^{'})$$

The resulting six clauses together are passed as the input to a SAT solver. Since the result is "Satisfiable" we know that our intermediate assignments are logically consistent with the circuit. On the other hand if our intermediate assignments were:

$$A = 1, B = 1 \text{ and } C = 0,$$

then we add the following clauses to the original CNF formula for the AND gate circuit:

$$(A)$$
$$(B)$$
$$(C^{'})$$

The resulting six clauses would return an "Unsatisfiable" result from the SAT solver, and hence we would know that the intermediate assignment is inconsistent with the original circuit.

We use a SAT solver instead of other possible options for detecting logical consistencies of the intermediate assignments due to the fact that generating the original CNF formula for the circuit needs to be done only once, we need only augment the golden (consistent) values in future calls. Also, Boolean satisfiability is a well-studied problem with highly efficient solvers such as BerkMin [11] easily available in the public domain.

4.6 Experimental Results

This section discusses two different sets of experiments. One set compares MLVC with other existing MLV determination techniques in terms of accuracy and run-times. The second set compares and discusses the mean and standard deviation of circuit leakage values computed by applying MLVC, MLVC-VAR and a random vector-based approach explained in the following subsection. All leakage values reported are in nA.

4.6.1 Selecting Parameter Values for MLVC and MLVC-VAR

For the results presented in this paper, we experimented with numerous combinations of the many parameters listed in Table 4.2. Against each parameter, the set of values considered during these experiments has also been listed.

We define a *method* as an assignment of values to each parameter within a set of parameters. The details of our experimentation for determining the parameter values chosen (or methods used) are explained next.

We choose the values of m_1 to be lower than m_2, so that we are more selective about states in the early iterations. m_3 is 1 since we accept all states in the final iteration. Values of β, γ and ρ are selected based on the scale of d_j, l_j and r_j. They are chosen such that a large value of β erases the effect of d_j on L or L_{var}, and a large value of γ erases the effect of l_j on L or L_{var}. A large value of ρ erases the effect of r_j on L_{var}. The various values of β, γ and ρ used are chosen such that our experiments explore the continuum along these three dimensions. $P_{\text{threshold}}$ and α need to be values close to 1 but not exactly 1, so we chose them to be 0.9 and 0.95, respectively.

For the MLVC approach, the parameters that can be varied are m_1, m_2, m_3, β, γ, $P_{\text{threshold}}$ and α. Therefore, the total number of methods can be 1,600. We ran ~19 benchmark circuits using these methods. The three methods that, among them, provided the best results for the maximum number of benchmark circuits (85%), were chosen for the rest of the MLVC experiments. These methods are called M1, M2 and M3, and their parameter assignments are listed in Table 4.3. Similarly, for

Table 4.2 Parameters' values considered in experiments for MLVC and MLVC-VAR

Parameter	Values				
m_1	0.4	0.5	0.6	0.7	
m_2	0.92	0.94	0.96	0.98	
m_3	1				
β	0.1	0	2	5	10
γ	10	20	50	70	100
ρ	0.2	0.5	1	2	
$P_{\text{threshold}}$	0.9	0.95			
α	0.9	0.95			

the MLVC-VAR approach, the parameters that can be varied are m_1, m_2, m_3, β, γ, ρ, $P_{\text{threshold}}$ and α. Therefore, the total number of methods we have are 6,400. Again, we ran ~19 benchmark circuits using these methods. The three methods that, among them, provided the best results for the maximum number of benchmark circuits were chosen for the rest of the MLVC-VAR experiments. These methods are called M1-Var, M2-Var and M3-Var, and their parameter assignments are listed in Table 4.7.

4.6.2 Comparing MLVC with Existing Techniques

We performed extensive experiments to validate MLVC and compare its results to the exact or near-exact minimum circuit leakage values. We created the leakage table for all gates in our library, i.e. computed the nominal leakage value for all input vectors, for all gates, using SPICE [16] with a 100-nm BPTM model card, at 30°C temperature. All our experiments were run on a 3.0-GHz Pentium 4 Linux machine with 1.0-GB RAM.

In all our experiments, we utilized a value of $k = 3$ iterations. The three methods (M1, M2 and M3) that we utilized for our experiments are described in Table 4.3. The value of p used was 1, but it can be increased for less accurate but faster invocations of the algorithm. The values reported in Table 4.3 were determined after extensive experimentation with many circuits as described in Sect. 4.6.1.

Methods M1 and M2 utilize a value of 0.6 for m_1. As a consequence, we expect to set more gates to platinum values in the first iteration. These methods are designed to reduce the number of gates discarded due to margin violations. Among these methods, M1 has a higher γ value, and therefore it biases the state selection towards states that have smaller distance. On the other hand, M2 has a higher β value, and as a consequence, state selection favors states with lower leakage. Method M3 has a smaller m_1 value, and therefore it tends to reject gates due to margin violations. It is biased towards state selections that have smaller distances. Our method exhibits very low runtimes. Given that the runtimes are very small, we can afford to apply all three methods (M1, M2 and M3) and choose the best result among the three. In general, we may try several methods and select one that yields the vector with the smallest leakage. In all experiments in this paper, we run each example using all available methods and then choose the best result.

In general the parameter sets need to be computed if the process technology is changed. This is done exactly once, hence it is a tractable task.

Table 4.3 Parameters used in our experiments for MLVC

Method	m_1	m_2	m_3	β	γ	$P_{\text{threshold}}$	α
M1	0.6	0.96	1	1	10	0.95	0.95
M2	0.6	0.96	1	0.1	100	0.95	0.95
M3	0.4	0.96	1	5	10	0.9	0.9

Table 4.4 Exhaustive and estimated leakages for small circuits

Circuit	Low	High	MLVC Low	R	R_{std}	Meth.	Time (s)
decod	78.29	122.67	78.29	0.00	0.00	M1	0
cm82a	115.20	133.00	115.20	0.00	0.00	M3	0.02
cm42a	106.64	141.87	115.38	0.25	0.08	M1	0.04
cm152a	80.10	124.64	84.35	0.10	0.05	M1	0.02
cm151a	93.48	141.83	103.08	0.20	0.10	M2	0.01
cm138a	98.19	136.22	98.19	0.00	0.00	M1	0.01
C17	19.76	37.99	20.07	0.02	0.02	M1	0.01
majority	36.69	57.40	40.51	0.18	0.10	M1	0
cm85a	183.23	271.51	221.16	0.43	0.21	M1	0.05
Avg				0.131	0.062		

Using these three methods, we first compared the results of MLVC with the exact minimum circuit leakages. This was performed for small examples, and results are reported in Table 4.4. The minimum leakage value returned by MLVC (Column 4), along with the exact maximum (Column 3) and minimum (Column 2) leakages are shown in this table. Further, we report a figure of merit R in Column 5.

$$R = \frac{\text{MLVC min leakage} - \text{Exact min leakage}}{\text{Exact max leakage} - \text{Exact min leakage}}.$$

The values of the maximum and minimum leakages are computed based on an exhaustive simulation of the circuit. Ideally, R should be 0. Runtimes for MLVC are reported in Column 8, while the method utilized is reported in Column 7.

Note that the figure of merit R is a more rigorous metric for comparing the effectiveness of any MLV determination technique. In the prior approaches to the MLV determination problem, the figure of merit utilized was

$$R_{\text{std}} = \frac{\text{Heuristic min leakage} - \text{Exact min leakage}}{\text{Exact min leakage}}.$$

Based on Table 4.4, the average value of R for MLVC was 0.13. For MLVC, the average value of the previously utilized figure of merit is 0.06.

Table 4.4 shows that the runtimes for MLVC are very small, with a good figure of merit for the method. The runtimes reported here are on average 10$\times$ faster then those reported in [12]. The reason for this improvement is the modification in the approach while choosing the best state for a selected gate. In the current form of MLVC, we use a SAT solver to test if a particular state s can be applied to a gate G without requiring to unroll the assignments from previous passes. Only if the state s clears this test do we consider it as a candidate state and then proceed to find the implications of assigning s on the other gates in the circuit, and compute L. In the published approach [12], implications on the other gates in the circuit (due to assigning s) were generated without first testing for satisfiability on s. For certain test cases, the approach in [12] generated its implications and then detected unsatisfiability. This led to the additional runtime.

Table 4.5 Leakages for large circuits

Circuit	N. gts	N. inps.	N. outs.	Low	High	MLVC low	R	R_{std}	Meth.	Time (s)
tcon	41	17	16	174.82	211.84	173.10	−0.05	−0.01	M1	0.05
cm163a	50	16	5	154.30	245.48	167.95	0.15	0.09	M1	0.04
pm1	52	16	13	191.90	269.98	208.20	0.21	0.08	M1	0.04
cm162a	56	14	5	186.46	264.69	204.01	0.22	0.09	M3	0.07
cm150a	58	21	1	203.63	340.77	245.81	0.31	0.21	M1	0.08
cu	62	14	11	205.01	306.62	214.63	0.09	0.05	M3	0.07
cc	74	21	20	269.61	354.98	295.70	0.31	0.10	M1	0.05
parity	75	16	1	276.78	363.04	278.50	0.02	0.01	M3	0.09
pcle	78	19	9	261.60	376.85	269.75	0.07	0.03	M1	0.05
pcler8	102	27	17	385.74	507.22	401.69	0.13	0.04	M3	0.09
lal	109	26	19	399.16	534.47	416.90	0.13	0.04	M2	0.22
b9	119	41	21	398.49	600.45	403.94	0.03	0.01	M3	0.20
unreg	120	36	16	440.20	538.22	452.32	0.12	0.03	M1	0.17
comp	131	32	3	454.01	613.59	486.08	0.20	0.07	M1	0.24
count	132	35	16	491.94	655.46	530.06	0.23	0.08	M1	0.27
c8	138	28	18	532.14	652.65	535.09	0.02	0.01	M2	0.20
cht	198	47	36	772.27	965.94	753.75	−0.10	−0.02	M3	0.57
ttt2	213	24	21	809.02	983.85	821.68	0.07	0.02	M3	0.71
C432	237	36	7	874.29	1110.63	929.66	0.23	0.06	M3	0.95
i5	198	133	66	858.95	945.20	875.81	0.20	0.02	M1	0.61
i3	258	132	6	1135.17	1327.06	1127.74	−0.04	−0.01	M1	1.16
x1	305	51	35	1148.05	1384.54	1180.63	0.14	0.03	M3	1.47
example2	330	85	66	1193.91	1472.34	1170.95	−0.08	−0.02	M1	1.95
x4	455	94	71	1705.54	2096.10	1724.77	0.05	0.01	M3	3.72
C1908	565	33	25	2115.12	2345.72	2181.09	0.29	0.03	M1	4.28
C499	582	41	32	2092.65	2249.22	2106.21	0.09	0.01	M1	4.20
rot	711	135	107	2805.18	3145.62	2890.93	0.25	0.03	M3	9.74
apex6	794	135	99	2930.22	3407.96	2977.63	0.10	0.02	M1	13.51
x3	908	135	99	3377.40	3744.96	3370.34	−0.02	0.00	M3	16.53
C3540	1354	50	22	5179.71	5757.48	5236.28	0.10	0.01	M3	41.03
C5315	1963	178	123	7982.04	8569.38	8027.37	0.08	0.01	M3	131.90
C6288	3734	32	32	14416.17	16000.10	14733.79	0.20	0.02	M3	540.07
C7552	2729	207	108	10989.43	11586.96	11087.33	0.16	0.01	M2	254.74
AVG							0.119	0.035		

We also tested MLVC on larger circuits. The results of this experiment are shown in Table 4.5. Columns 2,3 and 4 list the number of gates, number of inputs and number of outputs, respectively, for each circuit in Column 1. Columns 5 through 11 in this table are similar to Columns 2 through 8 in Table 4.4, with the exception that exact leakage values are not computed in this table. Instead, the minimum and maximum leakage found over 10,000 random vectors is shown in Table 4.5. According to [13], this statistically yields a greater than 99% confidence that we will obtain a leakage vector that is 0.5% from the minimum. *This is referred to as the Random Vectors Approach (RVA).*

Table 4.5 shows that MLVC produces minimum leakage vectors with very low errors, with extremely small runtimes. From [10], for the previously reported methods of [10], [17] and [19], the average errors were 5.3%, 3.7% and 10.4%, respectively (using the R_{std}[3] metric, for which MLVC results in an error of 3.5%). Further, the runtimes for MLVC are significantly smaller than those of [17], which is the most accurate known method for MLV determination.

4.6.3 *Comparing MLVC-VAR with MLVC and RVA*

Since, to the best of the authors' knowledge, there is no other work to date that considers PVT variations in the determination of the MLV, we compare the performance of MLVC-VAR with MLVC and RVA. We compare the mean, μ, and standard deviation, σ, of the circuit leakage values computed by applying the input vectors determined by MLVC-VAR, MLVC and RVA.

For use in the MLVC-VAR approach, we created an extended leakage table (for each gate) that contains the variations in leakage values due to PVT variations. For generating this table, we ran Monte Carlo (MC) simulations in SPICE, using the random PVT variations reported in [6], for 30,000 samples. These variations were assumed to be random (uncorrelated). Hence, the r_i (leakage range) values of a gate g were fixed. If detailed spatial information of gates was available, the correlation of these variables could be determined, resulting in different r_i values for different instances of any gate g. This can be done as follows:

Under intra-die variations, the value of any parameter p located at (x,y) can be modeled as [7, 8]:

$$p = p_n + x \cdot (S_x) + y \cdot (S_y) + e,$$

where p_n is the nominal design parameter value at die location (0,0), and S_x and S_y are gradients of the parameter indicating the spatial variations of parameter along the x and y directions, respectively. The term e stands for the random intra-chip variation, and the vector of all random components across the chip has a correlated multivariate normal distribution due to spatial correlations in the intra-chip variation. This vector depends on the correlation matrix of the spatially correlated parameters. Effectively, for each type of parameter, a correlation matrix of size $n \times n$, where n is the number of grid regions, represents the spatial correlation. This matrix could be determined from data extracted from manufactured wafers or derived from the spatial correlation models such as the one presented in [2]. Further, the correlation between different types of parameters can be added in this correlation matrix. This can be done by decomposing the correlated parameters into an uncorrelated set using an orthogonal transformation such as the principal component analysis (PCA) technique or by constructing a covariance matrix for all correlated parameters.

---

[3] Although the R metric is more rigorous, our comparisons to existing approaches utilize the R_{std} metric since these approaches utilize the R_{std} metric.

Table 4.6 Parameter variations

Parameter	μ	σ
Channel length	0.1 μm	0.05 μm
Power supply	1.2 V	0.04 V
Threshold voltage PMOS	0.3030 V	0.0127 V
Threshold voltage NMOS	0.2607 V	0.0110 V
Temperature	30°C	1°C

Table 4.7 Parameters used in our experiments for MLVC-VAR

Method	m_1	m_2	m_3	β	γ	ρ	$P_{\text{threshold}}$	α
M1-Var	0.6	0.96	1	5	10	2	0.95	0.95
M2-Var	0.6	0.96	1	5	10	0.2	0.95	0.95
M3-Var	0.4	0.96	1	1	70	1	0.90	0.90

By using this model in the generation of the extended leakage table (the table that tabulates the nominal, mean and standard deviation for every input, for each instance of any gate g), our approach can account for spatial correlation of every parameter. The steps of our approach, namely the selection of the best candidate gate and the selection of the best state for that gate, are decided using the data in the new extended leakage table. Note that our current implementation does not account for spatial correlations or correlation between different types of parameters. The above discussion, however, explains the methodology in order to account for these correlations. Implementing this methodology is a possible future work.

The mean and standard deviation of the PVT variables are listed in Table 4.6. We generated the μ and σ for the leakage values for all states for all gates in the library using the variations shown in Table 4.6.

In the experiments in this subsection, the parameter values used for MLVC were identical to those in Table 4.3. For MLVC-VAR, the parameters used are listed in Table 4.7.

Again, these parameters were chosen after extensive experimentation on several circuits. The margins m_1, m_2 and m_3 and parameters $P_{\text{threshold}}$ and α chosen are identical to those described in Table 4.3. Among the three methods M1-Var, M2-Var and M3-Var, M2-Var has the lowest value of ρ and therefore biases the state selection towards states with lower range. M3-Var favors states with lower leakage (since it has the lowest value of γ) whereas M1-Var favors states with lower distance, in comparison with M2-Var and M3-Var.

Table 4.8 compares the MLVC-VAR with MLVC and RVA. The μ and σ of the circuit leakage values are computed with similar Monte Carlo experiments as described previously for generating the extended leakage table. We use the same set of circuits as those used in Table 4.5. These are listed in Column 1. Column 2 reports the method used for MLVC-VAR. Note that the method used for MLVC is as reported in Table 4.5 for these circuits.

Columns 3 and 4 report μ and σ of the circuit leakage values computed by applying MLVC-VAR. The time taken for generating the input vector using MLVC-VAR is reported in Column 5. These runtimes, as expected, are about equal to those

Table 4.8 Comparing MLVC-VAR, MLVC and RVA

Circuit	MLVC-VAR					% Improv. w.r.t. MLVC			% Improv w.r.t. RVA		
	Method	μ	σ	Time (s)	*diff*	μ	$\mu \cdot \sigma$	$\mu + 6\sigma$	μ	$\mu \cdot \sigma$	$\mu + 6\sigma$
tcon	M2-Var	270.77	67.33	0.09	×	0.72	2.87	1.59	1.38	3.19	1.65
cm163a	M2-Var	243.36	55.01	0.04	√	−1.63	−16.76	−8.86	−3.50	−4.89	−2.25
pm1	M2-Var	296.41	65.09	0.08	√	7.24	13.13	6.74	0.82	3.46	1.88
cm162a	M1-Var	290.61	59.13	0.12	√	9.50	22.42	12.19	−1.38	4.00	2.41
cm150a	M1-Var	268.29	50.92	0.05	√	27.99	50.68	29.91	6.06	12.47	6.47
cu	M2-Var	306.70	58.74	0.12	√	18.43	33.79	18.64	0.68	7.56	4.12
cc	M1-Var	423.63	79.37	0.17	√	11.55	19.69	10.32	1.05	3.56	1.84
parity	M3-Var	403.71	67.47	0.18	√	3.67	13.60	7.11	3.19	13.61	7.13
pcle	M1-Var	397.66	75.78	0.21	√	11.94	18.01	9.32	2.62	4.53	2.27
pcler8	M3-Var	606.54	105.54	0.31	×	3.92	4.89	2.46	2.92	6.01	3.05
lal	M1-Var	648.27	98.66	0.17	√	−5.41	−17.43	−8.19	−5.95	−12.80	−6.19
b9	M3-Var	598.68	91.92	0.4	×	2.62	5.37	2.72	−1.38	−5.65	−2.72
unreg	M2-Var	645.86	82.79	0.35	×	4.22	8.50	4.33	6.97	28.40	14.71
comp	M2-Var	759.37	97.35	0.26	√	−2.09	−3.63	−1.83	−10.88	−10.87	−5.87
count	M3-Var	751.68	97.04	0.45	√	5.95	5.79	3.37	−1.17	6.01	2.61
c8	M3-Var	812.82	105.88	0.41	√	3.03	1.86	1.21	−2.05	−2.69	−1.42
cht	M1-Var	1,165.31	134.85	0.98	×	1.45	4.28	2.04	2.00	0.40	0.54
ttt2	M1-Var	1,224.16	128.36	1.03	√	8.97	19.44	9.96	0.99	7.93	3.40
C432	M1-Var	1,325.58	129.86	1.51	√	2.81	4.53	2.43	1.82	10.31	4.46
i5	M2-Var	1,181.13	136.61	1.57	√	10.14	10.98	6.59	10.10	14.35	7.97
i3	M3-Var	1,489.03	139.11	1.42	√	8.50	8.67	5.68	13.76	25.37	13.66
x1	M3-Var	1,875.22	176.48	1.97	√	3.44	−2.47	0.20	−5.59	−10.60	−5.28
example2	M2-Var	1,693.13	159.78	2.27	√	7.22	13.27	6.97	8.06	12.22	6.81
x4	M1-Var	2,414.85	169.22	5.27	√	7.67	15.69	7.97	8.38	21.22	10.12
C1908	M3-Var	3,257.70	216.46	6.16	√	4.14	9.46	4.55	0.47	2.78	1.00
C499	M3-Var	3,225.41	206.44	6.07	√	3.58	9.39	4.27	1.18	5.65	2.13
rot	M1-Var	4,119.71	242.07	11.49	√	6.11	8.39	5.18	4.97	9.39	4.89
apex6	M3-Var	4,126.95	236.48	15.26	√	9.51	18.18	9.53	8.84	17.21	8.93
x3	M3-Var	5,002.58	255.71	22.05	√	2.26	3.41	2.01	3.86	7.07	3.74
C3540	M3-Var	7,925.77	358.34	46.69	√	7.51	9.47	6.41	0.54	−1.26	0.05
C5315	M3-Var	11,598.26	415.87	134.95	√	6.13	11.78	6.11	6.33	12.02	6.29
C6288	M3-Var	22,012.21	577.94	339.72	√	3.99	6.49	3.81	1.69	2.79	1.61
C7552	M3-Var	16,544.98	484.45	272.22	√	2.29	5.94	2.51	1.53	4.68	1.78
Avg						5.98	9.69	5.37	2.07	5.98	3.08

reported for MLVC in Table 4.5. Note that MC simulations are performed one time, upfront for each gate. Hence, runtimes for MC simulations are not added in Column 5. Column 6, titled *diff* shows a "√" for circuits in which MLVC-VAR yielded a different input vector with respect to MLVC. A "×," on the other hand, represents that both MLVC-VAR and MLVC returned the same vector. On average, for about 85% of benchmarks, MLVC-VAR returns a different best case input vector as compared to MLVC. This reiterates the notion that MLVC alone might possibly yield a vector for which the worst case ($\mu + 6\sigma$) leakage of the combinational circuit is higher than what MLVC-VAR would compute. The leakage values for MLVC-VAR and MLVC are slightly different for the "×" circuits, since unassigned inputs are randomly set before Monte Carlo simulations. The next three columns report the percentage improvement of MLVC-VAR over MLVC. The percentage improvement (decrease) in μ is shown in Column 7, the percentage improvement (decrease) in $\mu \cdot \sigma$ is shown in Column 8 and the percentage improvement (decrease) in $\mu + 6\sigma$ is shown in Column 9. On average these improvements are 5.98%, 9.69% and 5.37%, respectively.

Similarly, Columns 10, 11 and 12 report the percentage improvement of MLVC-VAR over RVA. The percentage improvement in μ, $\mu \cdot \sigma$ and $\mu + 6\sigma$ over RVA is 2.07%, 5.98% and 3.08%, respectively.

It is important to note that PVT variations can in general result in large leakage variations. However, since we are considering leakage variations during the sleep mode of operation, our temperature variation is considered to have a σ of 1°C. This results in lowered leakage variations as can be observed in Table 4.8, than one would intuitively guess.

4.7 Summary

In this chapter, we have described a probabilistic method, MLVC, to perform input vector assignment for leakage minimization in a combinational circuit. We start by computing signal probabilities throughout the circuit. These probabilities are used to guide the selection of the next gate to assign. The selected gate is the one with the probabilistic highest leakage value and the largest leakage range due to process variations. Once this gate is selected, it is assigned a state, again in a manner that probabilistically minimizes its leakage. The implications induced by such a state selection are computed. A satisfiability solver is invoked to validate the state selection before our algorithm commits to this assignment. The algorithm terminates when all inputs have been assigned or are implied.

The MLVC technique is fast, flexible and provides accurate results. On average, for small examples, MLVC found minimum leakage values that were 6.2% from the minimum circuit leakage. For larger examples, it was impractical to compute the minimum circuit leakage exactly. We computed our statistics on the basis of running 10,000 samples of circuit leakage computation. For these examples, MLVC produces leakage vectors with leakage within 3.5% from the minimum. The runtimes of MLVC are much lower than existing techniques that produce results of similar quality. Additionally, the effect of PVT variations can be easily incorporated into such a probabilistic formulation.

A variant of MLVC, termed MLVC-VAR, was also presented. MLVC-VAR takes into account the effect of variations in leakage values due to PVT variations. Including the effect of PVT variations for determining minimum leakage vector is important because of the strong dependence of leakage currents on power supply, threshold voltage and temperature. Further, MLVC-VAR can be modified to account for spatial correlation and correlation between different parameter types as described in Sect. 4.6.3. This modification is a possible future work. The comparison of the mean and standard deviations of the circuit leakages induced by the input vectors generated by MLVC-VAR, MLVC and RVA further proves the relevance of taking into account the PVT variations while determining the MLV. On average, MLVC-VAR reports a 9.69% (5.37%) improvement in final circuit leakage over MLVC with respect to $\mu \cdot \sigma$ ($\mu + 6\sigma$) with similar runtimes. The improvement over RVA is 5.98% (3.08%) with much lower runtimes.

References

1. Abdollahi, A., Fallah, F., Pedram, M.: Runtime mechanisms for leakage current reduction in CMOS VLSI circuits. In: Proc., Symposium on Low Power Electronics and Design, pp. 213–218 (2002)
2. Agarwal, A., Blaauw, D., Zolotov, V., Sundareswaran, S., Zhao, M., Gala, K., Panda, R.: Statistical Delay Computation Considering Spatial Correlations. In: ASPDAC: Proceedings of the 2003 Conference on Asia South Pacific Design Automation, pp. 271–276. ACM Press, New York, NY, USA (2003). DOI http://doi.acm.org/10.1145/1119772.1119825
3. Agarwal, A., Kang, K., Roy, K.: Accurate Estimation and Modeling of Total Chip Leakage Considering Inter- & Intra-die Process Variations. In: ICCAD '05: Proceedings of the 2005 IEEE/ACM International Conference on Computer-Aided design, pp. 736–741. IEEE Computer Society, Washington, DC, USA (2005)
4. Aloul, F., Hassoun, S., Sakallah, K., Blauuw, D.: Robust SAT-Based Search Algorithm for Leakage Power Reduction. In: Proc. Power and Timing Models and Simulation. Seville, Spain (2002)
5. Bhardwaj, S., Vrudhula, S.B.K.: Leakage Minimization of Nano-scale Circuits in the Presence of Systematic and Random Variations. In: Proceedings, 42nd Design Automation Conference, pp. 541–546 (2005)
6. Cao, Y., Hu, C., Kahng, A.B., Sylvester, D.: Improved Estimates of Process Variation Impact on Deep Submicron Circuit Performance. In: Unpublished (2006)
7. Chang, H., Sapatnekar, S.S.: Full-Chip Analysis of Leakage Power Under Process Variations, Including Spatial Correlations. In: DAC '05: Proceedings of the 42nd Annual Conference on Design Automation, pp. 523–528. ACM Press, New York, NY, USA (2005). DOI http://doi.acm.org/10.1145/1065579.1065716
8. Chang, H., Sapatnekar, S.S.: Statistical Timing Analysis Under Spatial Correlations. In: Computer-Aided Design of Integrated Circuits and Systems, IEEE Transactions on, vol. 24, pp. 1467– 1482 (2005)
9. Cook, S.: The Complexity of Theorem-Proving Procedures. In: Proceedings, Third ACM Symp. Theory of Computing, pp. 151–158 (1971)
10. Gao, F., Hayes, J.: Exact and Heuristic Approaches to Input Vector Control for Leakage Power Reduction. In: Proc. International Conference on Computer-Aided Design, pp. 527–532. San Jose, CA (2004)
11. Goldberg, E., Novikov, Y.: BerkMin: A Fast and Robust SAT-Solver. In: Proc., Design Automation and Test in Europe (DATE) Conference, pp. 142–149 (2002)
12. Gulati, K., Jayakumar, N., Khatri, S.P.: A Probabilistic Method to Determine the Minimum Leakage Vector for Combinational Designs. In: Proceedings, IEEE International Symposium on Circuits and Systems (ISCAS), Kos, Greece, pp. 2241–2244 (2006)

13. Halter, J., Najm, F.: A Gate-Level Leakage Power Reduction Method for Ultra Low Power CMOS Circuits. In: Proc. Custom Integrated Circuits Conference, pp. 475–478. Santa Clara, CA (1997)
14. Johnson, M., Somasekhar, D., Roy, K.: Models and Algorithms for Bounds on Leakage in CMOS Circuits. IEEE Transactions on Computer-Aided Design of Integrated Circuits and Systems **18**(6), 714–725 (1999)
15. Li, X., Le, J., Pileggi, L.T.: Projection-Based Statistical Analysis of Full-Chip Leakage Power with Non-Log-Normal Distributions. In: Proceedings, 43rd Design Automation Conference, pp. 103–108 (2006)
16. Nagel, L.: SPICE: A Computer Program to Simulate Computer Circuits. In: University of California, Berkeley UCB/ERL Memo M520 (1995)
17. Naidu, S., Jacobs, E.: Minimizing Stand-by Leakage Power in Static CMOS Circuits. In: Proc., Design Automation and Test in Europe (DATE) Conference, pp. 370–376 (2001)
18. Narendra, S., De, V., Borkar, S., Antoniadis, D., Chandrakasan, A.: Full-Chip Sub-threshold Leakage Power Prediction for sub-0.18μm cmos. In: Proceedings, International Symposim on Low Power Electronics and Design, pp. 19–23 (2002)
19. Rao, R., Liu, F., Burns, J., Brown, R.: A Heuristic to Determine Low Leakage Sleep State Vectors for CMOS Combinational Circuits. In: Proc. International Conference on Computer-Aided Design, pp. 689–692. San Jose, CA (2003)
20. Rao, R., Srivastava, A., Blaauw, D., Sylvester, D.: Statistical Estimation Leakage Currents Considering Inter-and Intra-die Process Variations. In: Proceedings, International Symposim on Low Power Electronics and Design, pp. 84–89 (2003)
21. Rao, R., Srivastava, A., Blaauw, D., Sylvester, D.: Statistical Estimation of Leakage Current Considering Inter- and Intra-die Process Variation. In: ISLPED '03: Proceedings of the 2003 International Symposium on Low Power Electronics and Design, pp. 84–89. ACM Press, New York, NY, USA (2003). DOI http://doi.acm.org/10.1145/871506.871530
22. Saluja, N.S., Khatri, S.P.: Efficient SAT-Based Combinational ATPG Using Multi-Level Don't-Cares. In: Proceedings, IEEE International Test Conference (2005)
23. Su, H., Liu, F., Devgan, A., Acar, E., Nassif, S.: Full-Chip Leakage Estimation Considering Power Supply and Temperature Variations. In: Proceedings, International Symposim on Low Power Electronics and Design, pp. 78–83 (2003)
24. Unni, T.A., Walker, D.M.H.: Model-Based i_{DDQ} Pass/Fail Limit Setting. In: IEEE International Workshop on I_{dqq} Testing, pp. 43–47 (1998)
25. Zhang, S., Wason, V., Banerjee, K.: A Probabilistic Framework to Estimate Full-Chips Sub-threhold Leakage Power Distribution Considering Within-die and Die-to-Die P-T-V Variations. In: Proceedings, International Symposim on Low Power Electronics and Design, pp. 156–161 (2004)
26. Zhanping, C., Johnson, M., Liqiong, W., Roy, W.: Estimation of Standby Leakage Power in CMOS Circuit Considering Accurate Modeling of Transistor Stacks. In: Proc. International Symposium on Low Power Electronics and Design, pp. 239–244. Monterey, CA (1998)

Chapter 5
The HL Approach: A Low-Leakage ASIC Design Methodology

5.1 Overview

One of the most popular ways of reducing leakage is through the use high-V_T power gating transistors (as in the MTCMOS technique [8, 13] mentioned in Chap. 2). The HL approach is a variant of this technique that uses these power gating transistors selectively. In the HL approach we first create two low-leakage variants of each cell in a standard-cell library. If the inputs of a cell during the standby mode of operation are such that the output has a high value, we minimize the leakage in the pull-down network. Similarly we minimize leakage in the pull-up network if the output has a low value. In this manner, two low-leakage variants of each standard cell are obtained. While technology mapping a circuit, we determine the particular variant to utilize in each instance, so as to minimize leakage of the final mapped design.

This chapter is organized as follows. The philosophy of the HL approach is explained in Sect. 5.2. Related previous work is discussed in Sect. 5.3. In Sect. 5.4, details of the HL approach are presented. In Sect. 5.5, we present experimental results that compare placed-and-routed area, leakage and delay of this new methodology against MTCMOS and a regular standard-cell-based design style. The results show that the HL approach has better speed and area characteristics than MTCMOS implementations. The leakage current for HL designs can be dramatically lower than the worst-case leakage of MTCMOS-based designs and two orders of magnitude lower than the leakage of traditional standard cells. An ASIC design implemented in MTCMOS would require the use of separate power and ground supplies for latches and combinational logic, while our methodology does away with such a requirement. Another advantage of our methodology is that the leakage is precisely estimable, in contrast with MTCMOS. The primary contribution of the work presented in this chapter is a new low-leakage design style for static CMOS designs.

In Sect. 5.6, we present some experiments that explore the feasibility of using gate length biasing (minor changes to the channel length of a transistor) instead of changing the V_T. In Sect. 5.7, we discuss techniques to reduce leakage in dynamic (domino logic) designs and a summary is presented in Sect. 5.8.

N. Jayakumar et al., *Minimizing and Exploiting Leakage in VLSI Design*,
DOI 10.1007/978-1-4419-0950-3_5, © Springer Science+Business Media, LLC 2010

5.2 Philosophy of the HL Approach

The leakage current for a PMOS or NMOS device corresponds to the I_{ds} of the device when the device is in the *cut-off* or *sub-threshold* region of operation. The expression for this current [1] is:

$$I_{\mathrm{ds}}^{\mathrm{sub}} = \frac{W}{L} I_{\mathrm{D0}} \mathrm{e}^{\frac{V_{\mathrm{gs}}-V_{\mathrm{T}}-V_{\mathrm{off}}}{n v_{\mathrm{t}}}} \left[1 - \mathrm{e}^{-\frac{V_{\mathrm{ds}}}{v_{\mathrm{t}}}}\right] \tag{5.1}$$

Here I_{D0} and V_{off} (typically $V_{\mathrm{off}} = -0.08V$) are constants, while v_t is the thermal voltage (26 mV at 300°K) and n is the sub-threshold swing parameter.

We note that I_{ds} increases exponentially with a decrease in V_{T}. This is why a reduction in supply voltage (which is accompanied by a reduction in threshold voltage) results in exponential increase in leakage.

Another observation that can be made from (5.1) is that I_{ds} is significantly larger when $V_{\mathrm{ds}} \gg n v_{\mathrm{t}}$. For typical devices, this is satisfied when $V_{\mathrm{ds}} \simeq$ VDD. The reason for this is not only that the last term of (5.1) is close to unity, but also that with a large value of V_{ds}, V_{T} would be lowered due to drain-induced barrier lowering – DIBL (V_{T} decreases approximately linearly with increasing V_{ds}) [1, 15]. Therefore, leakage reduction techniques should ensure that the supply voltage is not applied across a single device, as far as possible.

Our approach to leakage reduction attempts to ensure that the supply voltage is applied across more than one *turned-off* device and one of those devices is a high-V_{T} device. This is achieved by selectively introducing a high-V_{T} PMOS or NMOS supply gating device in either the pull-up network of a gate (if the output is low in standby) or the pull-down network of a gate (if the output is high in standby). By this design choice, we obtain standard cells with both *low and predictable* standby leakage currents, unlike MTCMOS-based approaches.

5.3 Related Previous Work

Previous design approaches have suggested the use of dual-threshold devices [8] in an MTCMOS configuration which MTCMOS utilizes NMOS and PMOS power supply gating devices. The authors propose a MTCMOS standby device sizing algorithm, which is based on mutually exclusive discharging of gates. This technique is hard to utilize for random logic circuits as opposed to the extremely regular circuits, which are used as illustrative examples in [8]. In [13], the authors describe an MTCMOS implementation of a PLL using a 0.5-μm process. In both these works, the problem of estimating the leakage of an MTCMOS design is not addressed. In practice, the leakage of such a design can vary widely and is hard to control or predict. The threshold voltage is modified by bulk bias (via body effect) and DIBL, which are determined in part by the voltages of the bulk/source and source/drain nodes. Since cell inputs and outputs as well as bulk nodes float in an MTCMOS design operating in standby mode, precise prediction or control of leakage is impossible in MTCMOS. Cell input and output voltages affect the leakage of a gate as seen in (5.1). The bulk voltage V_b affects V_T through body effect, and sub-threshold leakage has an exponential dependence of V_T as seen in (5.1). Hence, MTCMOS designs can have a large range of leakage currents, with little ability to predict or control the actual leakage current.

The threshold voltage of a device drops due to the DIBL (Drain Induced Barrier Lowering) effect when V_{ds} is large [15]. Hence, leakage can be limited by making sure that the V_{ds} across a turned-off device is limited. In [9], the authors present a technique that ensures that the entire supply voltage (VDD) is not applied across one device. They propose an MTCMOS-like leakage reduction approach, in which the MTCMOS sleep devices are connected in parallel with diodes. This helps ensure that the V_{ds} across the sleep devices is no greater than VDD $- 2V_D$, where V_D is the forward-biased diode voltage drop.

5.4 The HL Approach

Our goal is to design standard cells with *predictably* low leakage currents. To achieve this, we design two variants of each standard cell. The two variants of each standard cell are designated "H" and "L." If the inputs of a cell during the standby mode of operation are such that the output has a high value, we minimize the leakage in the pull-down network. So a *footer* device (a high-V_T NMOS with its gate connected to $\overline{standby}$) is used. We call such a cell the "H" variant of the standard cell. Similarly, if the inputs of a cell during the standby mode of operation are such that the output has a low value, we minimize the leakage in the pull-up network by adding a *header* device (a high-V_T PMOS with its gate connected *standby*), and call such a cell the "L" variant of the standard cell.

This exercise, when carried out for a NAND3 gate, yields the circuits shown in Fig. 5.1. Note that the MTCMOS circuit is also shown in this figure. Although the PMOS and NMOS supply gating devices [equivalently called *header* and *footer* devices (devices shown shaded in Fig. 5.1] are shown in the circuit for the MTCMOS design, such devices are in practice shared by all the standard cells of a larger circuit block.

In our design approach, we utilized the same base standard-cell library for all design styles. Our standard-cell library consisted of INVA, INVB, NAND2A, NAND2B, NAND3, NAND4, NOR2, NOR3, NOR4, AND2, AND3, AND4, OR2, OR3, OR4, AOI21, AOI22, OAI21 and OAI22 cells. We utilized the *bsim100* predictive 0.1-μm model cards [4]. The devices have a $V_T^N = 0.26\,V$ and $V_T^P = -0.30\,V$. The header and footer devices we utilized had $V_T^N = 0.46\,V$ and $V_T^P = -0.50\,V$. We sized the header and footer devices so that the worst-case output delay penalty over all gate input transitions was no larger than 15% as compared to the regular standard cell using low V_T transistors. In [13] too, the power supply gating transistors were sized such that their simulated delay penalties were no larger

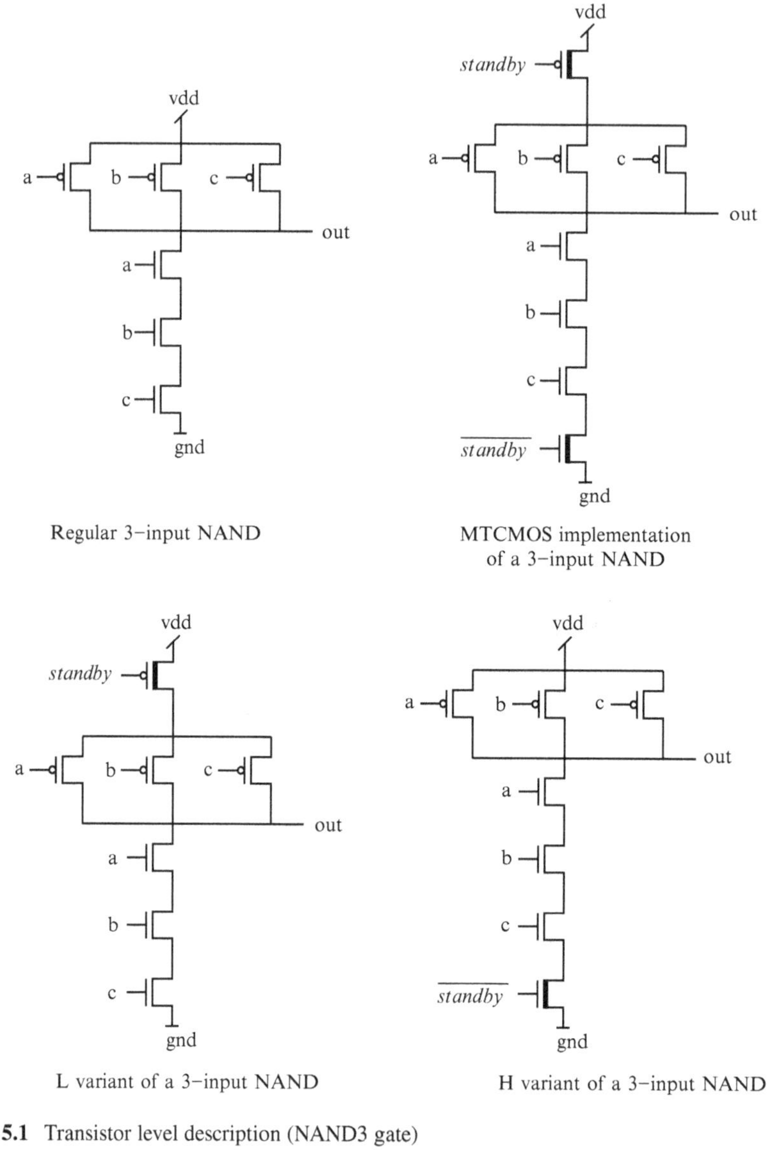

Fig. 5.1 Transistor level description (NAND3 gate)

than 15%. Additionally, if the delay penalty desired is less than 15%, then the gate area overheads are quite significant. The sizes of the devices of the regular standard cell were left unchanged in our MTCMOS and H/L cell variants.

If we were to modify the sizes of *all* devices of a gate (not just the header/footer devices), we anticipate that our cell area overheads would be much smaller, and the cells could be faster for a given area overhead. However, this would involve layout of H/L cells from scratch. For the results reported here, we have made a decision to

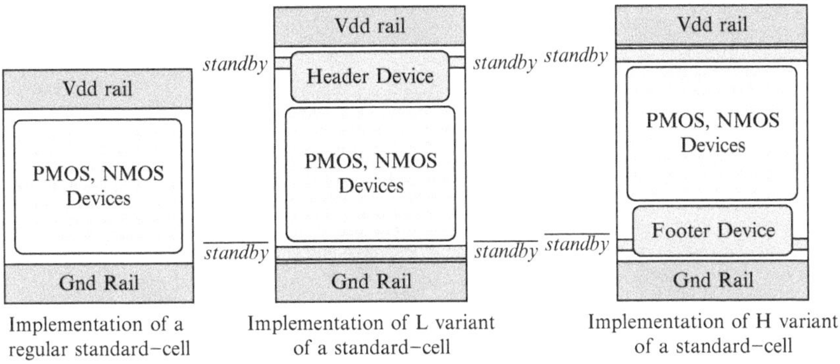

Fig. 5.2 Layout floor-plan of HL gates

not modify the device sizes of the regular design in order to produce an approach that is easy to adopt in practice. With this choice, we have been able to generate the layouts of the H/L standard cells by minimally modifying the layouts of the existing standard cells.

Our H/L cell layouts are derived from the existing standard cells by simply placing the VDD and GND rails of a cell further apart, in order to introduce just enough additional space to insert the header/footer devices. This is shown schematically in Fig. 5.2. Note that in the H and L variants of the regular standard cell, the layout of the regular standard-cell devices (the region labeled "PMOS, NMOS Devices") is not modified. The *standby* and $\overline{standby}$ signals are routed by abutment, and run across the width of each H/L standard cell. The header and footer transistors are implemented in a space-efficient zig-zag configuration as shown in the layout of Fig. 5.3. This also allows the header and footer device regions to be available for over-the-cell routing. In our simulations we assumed the width of the header and footer transistors to be equal to the center-line length of the poly shape. This is a common approximation used in circuit design. However, for additional accuracy one can conceivably run existing commercial extraction tools to obtain an adjustment factor to account for the U-turns made in the poly shape. However, the adjustment factor is expected to be close to unity since there are only two U-turns in each H and L cell. Finally our HL cells have more pin landing sites, to enable ease of routing. In this manner, we were able to design H/L layout variants of each cell in an area-efficient manner.

5.4.1 Design Methodology

The overall design flow to implement a circuit using H/L standard cells is very similar to a traditional standard-cell-based design methodology. We first perform traditional mapping using regular standard cells. After determining a set of primary input assignments for the standby mode of operation, we simulate the circuit with

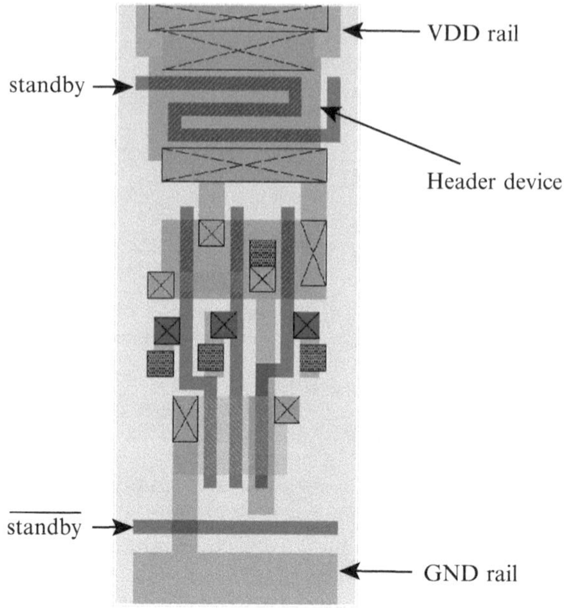

Fig. 5.3 Layout of NAND3-L cell

these assignments to determine the output of each gate. If the output of a gate is high, we replace it with the corresponding H cell, and if it is low, with the L variant of the cell. Hence, the decision of which cell variant to utilize for any given circuit can be made in time linear in the size of the circuit.

The schemes discussed in [6, 12, 17] are similar to ours, but their authors do not mention that the leakage current in such a scheme is predictable. Also, in our HL methodology, the power supply gating devices are included within the standard-cell itself for simplicity. This ensures that we do not have to use ungated additional power supply rails, which are required in the schemes of [6, 12, 17]. We also perform detailed analysis of the delay-area trade-offs for an extensive set of benchmark circuits, which is discussed in Sect. 5.5.

The determination of an optimal primary input assignment to utilize for the standby mode is an NP-hard problem. Chapters 3 and 4 provide some solutions to finding the optimal input vector. For a scan-enabled design, these primary inputs can be easily applied. If this is not the case, a phase-forcing circuit as discussed in [12] can be used to apply the required inputs to a combinational block.

5.4.2 Advantages and Disadvantages of the HL Approach

The advantages of the HL methodology are as follows:

- By ensuring that each cell has a full-rail output value during standby operation, we make sure that the leakage of each standard cell, and therefore the leakage of a standard-cell based design, is *precisely predictable*. Therefore, our methodology avoids the unpredictability of leakage that results when using the MTCMOS style of design. This unpredictability occurs due to the fact that in MTCMOS, cell outputs, inputs and bulk voltages float to unknown values that are dependent on various processing and design factors.

- Since our inverting H/L cells utilize exactly one supply gating device (as opposed to two devices for MTCMOS), our cells exhibit better delay characteristics than MTCMOS for one output transition (the falling transition for L gates and rising transition for H gates). Though the authors of [13] mention that it possible to use only footer devices, their implementation uses both header and footer devices. Though using only a footer device will reduce the delay penalties, the leakage current increases as we show in Sect. 5.5.
- For MTCMOS designs, memory elements would require clean power and ground supplies if they were to retain state during standby mode [13]. With the HL approach, the inputs to a combinational block are fixed in the standby mode. Hence, the states of the memory elements that drive these inputs are also fixed. Therefore, our technique can be applied to sequential elements as well (by using header devices when the leakage path is through the PMOS stack and using footer devices when the leakage path is through the NMOS stack). Alternatively, we could utilize the same flip-flop design as in [13]. In either case, the HL approach would *not* require special clean supplies to be routed to the flip-flop cell, resulting in lower area utilization for sequential designs.
- For many of the standard cells, and particularly for larger cells that exhibit large values of leakage, our H/L cells exhibit much lower leakage current. However, there are cells for which our cells exhibit comparable or greater leakage than MTCMOS as well. This is quantified in Sect. 5.5.
- By implementing the header and footer devices in a layout-efficient manner, we ensure that the layout overhead of H/L standard-cells is minimized. Our choice of layout also allows the header and footer device regions to be free for over-the-cell routing.

The disadvantages of our approach are as follows:

- The determination of the primary input assignments to utilize for the standby mode is a complex once. Although our current implementation makes this decision arbitrarily, it can be improved by applying the ideas described in Sect. 5.4.1.
- Using the HL approach requires that the primary inputs to the circuit be driven to known values in the *standby* state. However, if we assume that a combinational block of logic implemented using our approach is driven by flip-flops that are scan-enabled, then the required input vector can be simply scanned in before the circuit goes into the *standby* state. Alternatively, special circuitry (such as a NAND2 or a NOR2 gate with the standby signal as one of the inputs) could be added at the primary inputs.
- The experiments presented in this chapter can be improved if the technology mapping tools are modified. Assuming that the primary input vector is predetermined and that we use a dynamic programming-based technology mapper, the

mapper would need to store the best match at any node as well as the logic state of that best match. For any new node that is being mapped, its logic state can therefore be determined, and so we would know whether to use a H or L cell for that mapping. In either case, we would know what delay or area value to use for an optimum match at that node. In reality, the problems of technology mapping and the determination of an optimal primary input vector are coupled.
- Our method requires that the standby signals be routed to each cell. However, we have overcome this problem by designing the layout of H/L cells such that the routing of standby signal is performed by abutment, while also leaving free space for over-the-cell routing above the region where the standby signals are run.

5.5 Experimental Results

The standards cells we used were taken from the low-power standard-cell library of [2]. Our standard-cell library consisted of the following cells: INVA, INVB, NAND2A, NAND2B, NAND3, NAND4, NOR2, NOR3, NOR4, AND2, AND3, AND4, OR2, OR3, OR4, AOI21, AOI22, OAI21 and OAI22. The H and L variants of each of the standard cells were created by modifying (adding high-V_T header and/or footer devices as required) the regular cells. The header and footer devices used in the HL variants as well as the MTCMOS cells were sized such that the worst-case cell delays were within 15% of the regular standard-cell worst-case delays. The sizes of the other transistors were not changed for reasons mentioned in Sect. 5.4.

We used SPICE3f5 [14] for simulations of the standard cells. The NMOS and PMOS model cards used were derived from the *bsim100* model cards [4]. The threshold voltages of the high-V_T transistors were 200 mV greater than those of the regular devices. A supply voltage of 1.2 V was assumed.

After performing the design, layout and characterization of individual cells, we compared the leakage, delay and area characteristics of the HL, MTCMOS and regular standard-cell-based design methodologies for a set of circuits taken from the MCNC91 benchmark suite.

In Fig. 5.4, we plot the range of leakage values for each MTCMOS cell against the range of leakage values obtained using the corresponding HL cell. For the HL cells, all possible input vectors were applied for each cell. This gave us the range of leakage values possible for the HL cells. Finding the range of leakage for the MTCMOS cells, is not as straightforward as finding the leakage for HL cells, since the inputs to the MTCMOS cell are not full-rail values during standby. For our experiments, we applied all possible voltage values from 0 to 1.2 V, in steps of 0.2 V, at each input of the MTCMOS cells and then found the minimum and maximum leakage currents. Note that in Fig. 5.4, we have also compared the range of leakage values for MTCMOS cells using only header sleep transistors and for MTCMOS cells using only footer sleep transistors. From Fig. 5.4, we find that the range of leakage values for the MTCMOS cells using both header and footer sleep transistors

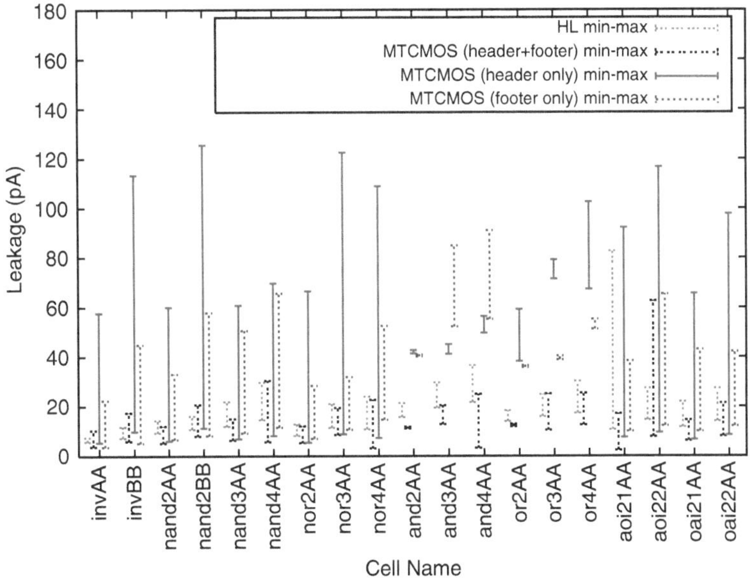

Fig. 5.4 Plot of leakage range of HL vs. MT method

is much smaller than the range of leakage values when only one of the devices is used. Hence, from this point on, we use only the MTCMOS cells with both header and footer devices for comparisons with our H/L cells.

5.5.1 Comparison of Placed and Routed Circuits

A set of circuits from the MCNC91 benchmarks were implemented using all three design methodologies (regular standard-cell, HL and MTCMOS). Logic optimization and mapping were performed in the SIS [16] environment. The resulting leakage, area and delay numbers were compared. For circuits designed using H/L type cells, each primary input signal was assumed to be logic low in standby mode. The choice of selecting the H or L variant for each standard cell was made as described in Sect. 5.4.1.

5.5.1.1 Leakage Comparison

We first computed the leakage of each H/L cell based on the values of cell inputs implied by the applied primary input combination. Using this information, the leakage of the circuit mapped using the H/L gates was estimated by adding the leakage of the individual gates used. This is possible since the inputs to each gate in standby

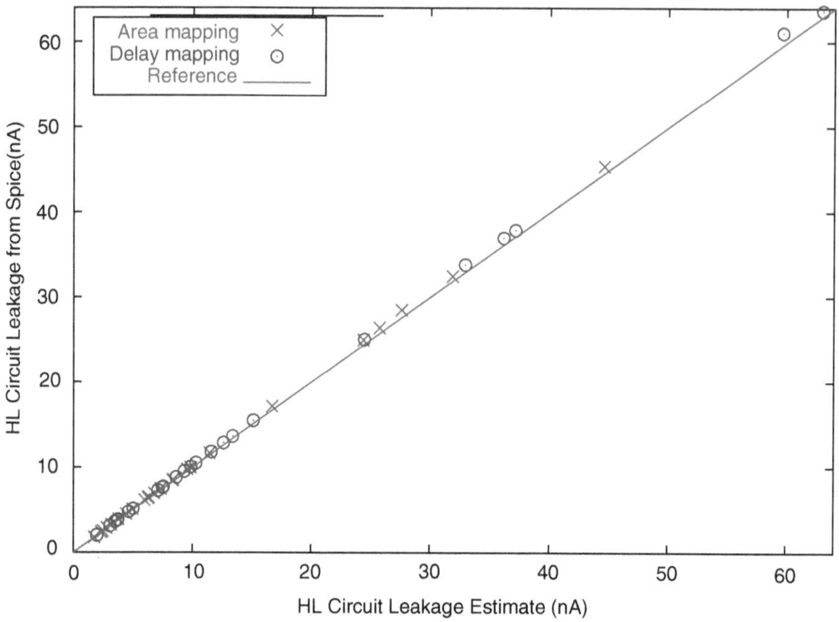

Fig. 5.5 Leakage of HL-spice vs. HL method over circuits

mode are known. We also ran SPICE on the mapped design, using the same primary input vector, to obtain a more accurate leakage estimate for the design. Figure 5.5 is a scatter plot of the leakage values thus obtained, for all the circuits under consideration. From Fig. 5.5, we observe that for all the examples, the estimated leakage for the HL design and actual leakage obtained from SPICE are in very close agreement. This validates our claim that the *leakage for a HL design is precisely estimable* from the leakage values of each of its constituent gates. Thus, if one were to design low-leakage circuitry using the HL methodology, the standby power consumption can be computed with great accuracy. This is in stark contrast with MTCMOS-based designs.

For the MTCMOS methodology, we determined the sum of the maximum and minimum leakage values of individual gates (these values were also previously estimated from SPICE simulations and reported in Fig. 5.4). The results are presented in Figs. 5.6 and 5.7 and compared with the leakage of the HL methodology. In Fig. 5.6, the circuits were mapped for minimum area, while in Fig. 5.7, the circuits were mapped for minimum delay. In a mapped design, the inputs to the MTCMOS gates of the circuit would float in standby mode. Therefore, the precise leakage value for the MTCMOS design is unpredictable, hence we used the maximum and minimum values of MTCMOS leakage as mentioned in the description for Fig. 5.4. In practice, the actual value of the leakage current for an MTCMOS circuit may well be greater than the maximum value as computed above, based on the voltage values of the gate inputs and bulk nodes that float during standby.

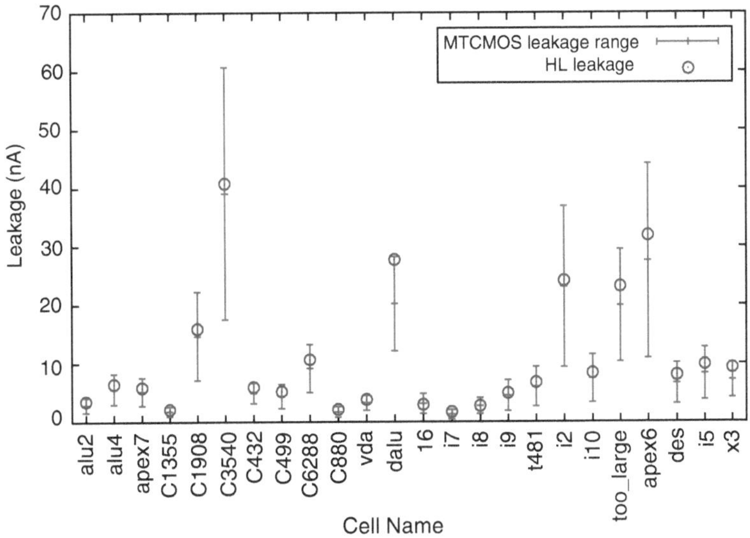

Fig. 5.6 Leakage of HL vs. MT (circuits mapped for min. area)

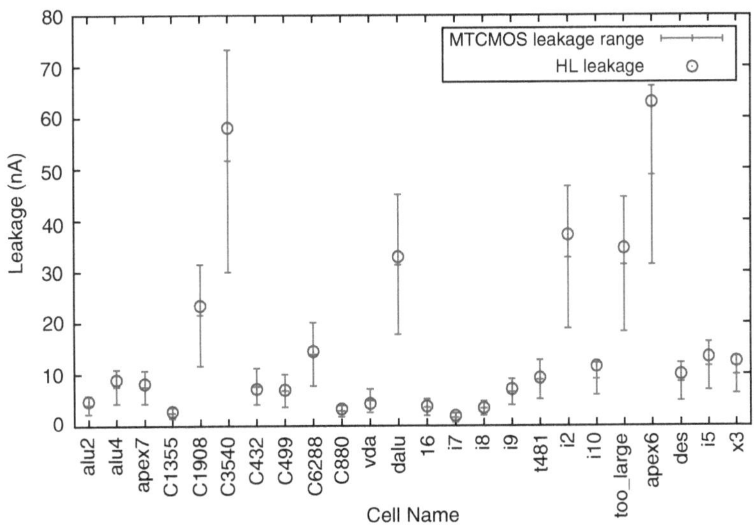

Fig. 5.7 Leakage of HL vs. MT (circuits mapped for min. delay)

Figures 5.6 and 5.7 indicate that the leakage of a design implemented using HL cells can be much smaller than the maximum leakage of a MTCMOS design. Note that for the results presented here, we simply assumed that the primary inputs were set to logic 0. If we were to set the primary input vector to a state that minimized leakage, the leakage for our approach is expected to be even lower.

5.5.1.2 Delay Comparison

To compare the delay of the three techniques, we performed exact timing analysis [11]. Given a mapped circuit, exact timing analysis returns the largest *sensitizable* delay for that circuit. As opposed to static timing analysis, exact timing eliminates false paths. We used the implementation of exact timing (the *sense* package that is implemented in SIS [16]) from the authors of [11].

To run *sense*, we generated a modified library description file for each of the three techniques. This file, in SIS's *genlib* format, describes the rising and falling delay from each input pin to the output pin for all gates in the library. Each such delay is a tuple consisting of a constant delay term and a load-dependent term. A standard-cell library characterization script was utilized to automatically generate this *genlib* file for all three design styles.

The results of *sense* are described in Table 5.1 (for the case where mapping is done for delay minimization) and Table 5.2 (for the case where mapping is done for area minimization). For our benchmark suite of 24 examples, HL mapping exhibits a delay overhead of about 10% while MTCMOS exhibits an area overhead of 12.5%, compared to the regular method. As discussed earlier, the delay of the HL circuit

Table 5.1 Delay (ps) comparison for all methods (delay mapping)

Example	Reg delay	HL delay	HL ovh.	MT delay	MT ovh.
alu2	4,146.65	4,296.20	3.61	4,546.15	9.63
alu4	5,024.59	5,135.15	2.20	5,583.55	11.12
apex6	1,660.15	1,644.10	−0.97	1,754.70	5.70
apex7	1,959.00	1,916.60	−2.16	2,108.40	7.63
dalu	9,270.03	10,314.05	11.26	10,494.15	13.21
des	14,571.29	16,690.05	14.54	16,704.20	14.64
C1355	2,567.91	2,738.10	6.63	2,922.80	13.82
C1908	3,056.04	3,403.45	11.37	3,467.75	13.47
C3540	5,756.18	6,577.75	14.27	6,537.05	13.57
C432	5,309.39	5,679.95	6.98	6,015.25	13.29
C499	2,289.99	2,439.05	6.51	2,586.20	12.93
C6288	13,632.70	15,528.65	13.91	15,742.70	15.48
C880	2,509.65	2,853.90	13.72	2,890.80	15.19
i2	610.55	652.70	6.90	665.95	9.07
i5	1,136.75	1,225.45	7.80	1,232.35	8.41
i6	6,698.08	7,598.70	13.45	7,610.40	13.62
i7	8,074.18	9,162.45	13.48	9,174.15	13.62
i8	19,027.58	21,498.20	12.98	21,799.45	14.57
i9	7,370.84	8,475.55	14.99	8,503.00	15.36
i10	8,479.30	8,850.95	4.38	9,680.85	14.17
t481	10,040.29	11,398.90	13.53	11,374.05	13.28
too_large	4,407.89	4,809.00	9.10	4,998.65	13.40
vda	3,890.79	4,329.05	11.26	4,439.20	14.10
x3	2,363.04	2,653.60	12.30	2,680.30	13.43
Avg			9.25%		12.61%

Table 5.2 Delay (ps) comparison for all methods (area mapping)

Ckt.	Reg. delay	HL delay	HL ovh.	MT delay	MT ovh
alu2	3,971.00	4,285.60	7.92	4,474.70	12.68
alu4	6,068.20	6,797.55	12.02	6,909.25	13.86
apex6	2,248.85	2,530.45	12.52	2,500.20	11.18
apex7	1,871.10	1,925.60	2.91	2,037.95	8.92
dalu	11,868.45	12,807.75	7.91	13,198.00	11.20
des	19,564.60	20,593.90	5.26	22,228.00	13.61
C1355	2,952.80	3,232.40	9.47	3,383.60	14.59
C1908	4,087.80	4,689.80	14.73	4,676.70	14.41
C3540	5,730.85	6,258.55	9.21	6,528.40	13.92
C432	5,220.30	5,638.00	8.00	5,893.10	12.89
C499	2,723.60	3,053.90	12.13	3,117.60	14.47
C6288	11,352.30	12,912.65	13.74	13,151.30	15.85
C880	2,685.50	2,963.30	10.34	2,995.70	11.55
i2	703.00	763.60	8.62	787.60	12.03
i5	1,154.70	1,287.30	11.48	1,270.80	10.05
i6	9,182.30	10,564.60	15.05	10,409.20	13.36
i7	10,549.85	11,944.90	13.22	11,781.10	11.67
i8	24,974.05	28,940.35	15.88	28,675.30	14.82
i9	14,746.35	16,497.85	11.88	16,576.30	12.41
i10	10,335.00	11,532.15	11.58	11,664.95	12.87
t481	17,192.70	19,317.20	12.36	19,092.50	11.05
too_large	4,205.35	4,650.85	10.59	4,647.90	10.52
vda	5,465.45	6,140.05	12.34	6,170.55	12.90
x3	3,591.25	3,986.60	11.01	3,915.80	9.04
Avg			10.84%		12.49%

is lower on account of the fact that only one transition of each gate is degraded in the process of modifying a gate for reduced leakage in the H/L approach. We also find that in two cases (apex7 and apex6 in Table 5.1), the HL circuit actually has a small delay decrease. This is due to the fact that while adding a footer sleep device worsens the falling transition, the rising transition actually improves slightly. This is because the additional footer sleep device makes the path to ground more resistive and hence speeds up the rising transition. Similarly, falling transitions are improved slightly when a header sleep device is used. Hence, in rare cases it is possible that a critical path gets sped up due to the addition of sleep transistors.

5.5.1.3 Area Comparison

We optimized and mapped our benchmark designs (for both minimum area and minimum delay) using SIS [16]. The circuits were then placed and routed using the Silicon Ensemble [3] tool set from Cadence Design Systems. Placement and routing was performed for both regular standard-cell and H/L cell-based circuits, using 2, 3 and 4 metal routing layers. This gave us an accurate measure of the actual die area required to design circuits using these two methodologies. For the MTCMOS methodology, the header and footer "sleep" transistors are large devices, which are shared by all the gates in a design. According to [8], one can exploit information about simultaneous transitions in a circuit to size sleep transistors efficiently. As stated earlier, this approach is not feasible for random logic circuits. Therefore, for MTCMOS circuits, we found the sum of the sizes of the MTCMOS headers and footers of the individual gates in the design. Based on this information, we estimated the layout area overhead of MTCMOS. This overhead was then added to the area of the circuit implemented using regular cells. In an MTCMOS design, additional area needs to be devoted for routing an extra pair of power rails (see Sect. 5.4.2). This was neglected since our designs were combinational in nature. For sequential circuits, the MTCMOS overhead would therefore be higher. Tables 5.3 and 5.4 describe the area comparison results. The former table is obtained when technology mapping was performed for minimum delay, and the latter for minimum area. The tables show the total area (using a 0.1 μ process) for regular standard cell, HL cell and MTCMOS-based circuits. The percentage area overhead for the HL and MTCMOS methods is also shown.

We note that on average, the HL design methodology exhibits a 11–30% area overhead compared to the regular design. However, the HL designs utilize on average up to 17% less area than the MTCMOS designs. As seen in Tables 5.3 and 5.4, the area overhead for MTCMOS does not decrease with increased metal layers, while the area overhead for HL does decrease. This is because the distributed nature of the sleep transistors in the HL scheme allows for more over-the-cell routing opportunities. The results validate the intuition that when more metal layers are used, the router can take advantage of over-the-cell routing and the area penalty for the HL methodology is reduced. For some examples, the HL designs exhibit a lower area than their regular counterparts. We conjecture that this is due to the fact that our HL cells are more router-friendly, with more over-the-cell routing space and also more pin landing sites.

5.6 Using Gate Length Biasing Instead of V_T Change

Recent research [5,7] has suggested that *gate-length biasing* can be used alternative to multiple threshold voltage devices. Gate-length biasing is a technique by which small increases (5–10%) in the gate length are made, reducing leakage by as much as 2×. Gate-length biasing does not require additional lithography masks and is hence inexpensive to implement. We replaced the high-V_T devices in the H/L cells with devices with longer channel length (and low V_T) in an effort to see how this would affect the delay and leakage of the H/L cells. We tried gate lengths that were 10% higher and 20% higher than nominal (100nm). The minimum and maximum leakages obtained for each of the cells are shown in Fig. 5.8.

Note that in Fig. 5.8, the leakage of the regular H/L cells (that use high-V_T header or footer transistors) has been multiplied by a factor of 10, while the leakage of the

Table 5.3 Area (μ^2) comparison for all methods (delay mapping)

	2-Layer						4-Layer					
Ckt.	Reg. area	HL area	HL ovh.	MT Area	MT ovh.	HL-MTovh.	Reg. Area	HL Area	HL ovh.	MT Area	MT ovh.	HL-MTovh.
alu2	2,480.04	3,203.56	29.17	3,422.48	38.00	−6.40	1,713.96	2,560.36	49.38	2,656.40	54.99	−3.62
alu4	5,184.00	6,400.00	23.46	6,964.54	34.35	−8.11	3,576.04	4,542.76	27.03	5,356.58	49.79	−15.19
apex6	4,928.04	5,565.16	12.93	6,740.40	36.78	−17.44	4,070.44	4,542.76	11.60	5,882.80	44.52	−22.78
apex7	1,156.00	1,600.00	38.41	1,756.09	51.91	−8.89	1,089.00	1,459.24	34.00	1,689.09	55.10	−13.61
dalu	10,816.00	14,352.04	32.69	15,509.27	43.39	−7.46	9,101.16	12,678.76	39.31	13,794.43	51.57	−8.09
des	46,397.16	51,710.76	11.45	56,678.33	22.16	−8.76	48,664.36	28,425.96	−41.59	58,945.53	21.13	−51.78
C1355	3,203.56	4,542.76	41.80	5,059.68	57.94	−10.22	3,672.36	4,542.76	23.70	5,528.48	50.54	−17.83
C1908	3,387.24	4,761.00	40.56	4,912.76	45.04	−3.09	3,249.00	3,969.00	22.16	4,774.52	46.95	−16.87
C3540	7,744.00	10,120.36	30.69	10,871.04	40.38	−6.91	5,806.44	7,779.24	33.98	8,933.48	53.85	−12.92
C432	1,169.64	1,747.24	49.38	1,819.41	55.55	−3.97	1,197.16	1,681.00	40.42	1,846.93	54.28	−8.98
C499	2,134.44	3,069.16	43.79	3,252.29	52.37	−5.63	3,624.04	2,704.00	−25.39	4,741.89	30.85	−42.98
C6288	13,041.64	17,476.84	34.01	21,098.58	61.78	−17.17	11,620.84	16,952.04	45.88	19,677.78	69.33	−13.85
C880	1,814.76	2,480.04	36.66	2,586.02	42.50	−4.10	1,428.84	2,134.44	49.38	2,200.10	53.98	−2.98
i2	1,024.00	1,398.76	36.60	1,358.97	32.71	2.93	817.96	1,142.44	39.67	1,152.93	40.95	−0.91
i5	1,918.44	2,560.36	33.46	2,676.54	39.52	−4.34	2,916.00	2,116.00	−27.43	3,674.10	26.00	−42.41
i6	3,576.04	4,705.96	31.60	4,999.98	39.82	−5.88	4,070.44	3,969.00	−2.49	5,494.38	34.98	−27.76
i7	6,177.96	6,115.24	−1.02	8,054.29	30.37	−24.07	4,070.44	5,212.84	28.07	5,946.77	46.10	−12.34
i8	20,449.00	26,830.44	31.21	27,105.78	32.55	−1.02	21,609.00	20,449.00	−5.37	28,265.78	30.81	−27.65
i9	5,184.00	6,561.00	26.56	6,824.72	31.65	−3.86	4,019.56	5,745.64	42.94	5,660.28	40.82	1.51
i10	28,968.04	30,765.16	6.20	35,796.49	23.57	−14.06	24,649.00	18,117.16	−26.50	31,477.45	27.70	−42.44
t481	24,964.00	33,489.00	34.15	33,259.85	33.23	0.69	20,334.76	29,104.36	43.13	28,630.61	40.80	1.65
too_large	5,685.16	7,396.00	30.09	7,456.22	31.15	−0.81	3,769.96	5,270.76	39.81	5,541.02	46.98	−4.88
vda	7,992.36	10,000.00	25.12	10,111.07	26.51	−1.10	4,928.04	6,822.76	38.45	7,046.75	42.99	−3.18
x3	6,304.36	7,499.56	18.96	8,285.65	31.43	−9.49	4,928.04	5,745.64	16.59	6,909.33	40.20	−16.84
AVG			29.08		38.94	−7.05			20.70		43.97	−16.95

Table 5.4 Area (μ^2) comparison for all methods (area mapping)

	2-Layer						4-Layer					
Ckt.	Reg. area	HL area	HL ovh.	MT Area	MT ovh.	HL-MTovh.	Reg. Area	HL Area	HL ovh.	MT Area	MT ovh.	HL-MTovh.
alu2	2,097.64	2,560.36	22.06	2,626.41	25.21	−2.51	1,296.00	1,764.00	36.11	1,824.77	40.80	−3.33
alu4	4,356.00	5,685.16	30.51	5,343.56	22.67	6.39	2,601.00	3,528.36	35.65	3,588.56	37.97	−1.68
apex6	3,721.00	4,435.56	19.20	4,667.28	25.43	−4.96	4,542.76	3,113.64	−31.46	5,489.04	20.83	−43.28
apex7	912.04	1,296.00	42.10	1,257.38	37.86	3.07	795.24	1,142.44	43.66	1,140.58	43.43	0.16
C1355	2,323.24	3,433.96	47.81	3,324.24	43.09	3.30	2,209.00	2,981.16	34.96	3,210.00	45.31	−7.13
C1908	2,601.00	3,624.04	39.33	3,417.87	31.41	6.03	2,894.44	2,601.00	−10.14	3,711.31	28.22	−29.92
C3540	6,241.00	8,281.00	32.69	7,844.76	25.70	5.56	4,489.00	5,745.64	27.99	6,092.76	35.73	−5.70
C432	817.96	1,156.00	41.33	1,116.43	36.49	3.54	729.00	1,011.24	38.72	1,027.47	40.94	−1.58
C499	1,764.00	2,480.04	40.59	2,381.99	35.03	4.12	1,521.00	2,135.36	40.39	2,138.99	40.63	−0.17
C6288	10,774.44	15,525.16	44.09	15,035.06	39.54	3.26	9,025.00	12,056.04	33.58	13,285.62	47.21	−9.25
C880	1,369.00	1,989.16	45.30	1,859.69	35.84	6.96	1,197.16	1,648.36	37.69	1,687.85	40.99	−2.34
dalu	9,254.44	11,793.96	27.44	11,834.39	27.88	−0.34	6,304.36	8,353.96	32.51	8,884.31	40.92	−5.97
des	45,710.44	47,089.00	3.02	51,786.20	13.29	−9.07	51,892.84	22,560.04	−56.53	57,968.60	11.71	−61.08
i2	772.84	1,142.44	47.82	1,041.92	34.82	9.65	817.96	985.96	20.54	1,087.04	32.90	−9.30
i5	1,681.00	2,246.76	33.66	2,210.91	31.52	1.62	1,197.16	1,600.00	33.65	1,727.07	44.26	−7.36
i6	3,433.96	3,069.16	−10.62	4,172.76	21.51	−26.45	4,070.44	2,560.36	−37.10	4,809.24	18.15	−46.76
i7	4,928.04	5,184.00	5.19	5,868.12	19.08	−11.66	3,624.04	3,203.56	−11.60	4,564.12	25.94	−29.81
i8	17,902.44	19,656.04	9.80	21,382.48	19.44	−8.07	18,769.00	12,588.84	−32.93	22,249.04	18.54	−43.42
i9	4,329.64	4,928.04	13.82	5,415.02	25.07	−8.99	4,070.44	3,969.00	−2.49	5,155.82	26.67	−23.02
i10	28,968.04	29,584.00	2.13	32,519.46	12.26	−9.03	21,609.00	13,409.64	−37.94	25,160.42	16.43	−46.70
t481	20,107.24	25,027.24	24.47	24,616.10	22.42	1.67	12,321.00	17,056.36	38.43	16,829.86	36.59	1.35
too_large	5,155.24	6,432.04	24.77	6,232.95	20.91	3.19	3,249.00	4,019.56	23.72	4,326.71	33.17	−7.10
vda	7,022.44	8,427.24	20.00	8,139.24	15.90	3.54	4,225.00	5,329.00	26.13	5,341.80	26.43	−0.24
x3	5,041.00	6,822.76	35.35	6,400.07	26.96	6.60	5,929.00	4,542.76	−23.38	7,288.07	22.92	−37.67
AVG			26.74		27.06	−0.52			10.84		32.36	−17.55

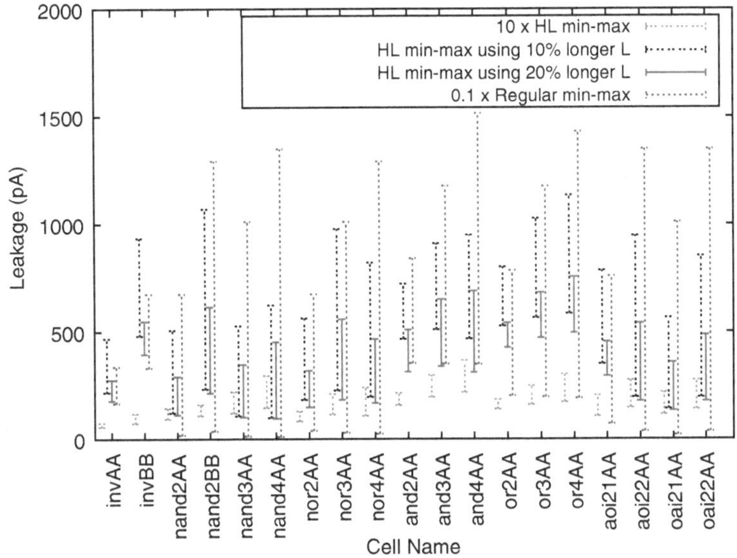

Fig. 5.8 Plot of leakage range of H/L cells, H/L cells with gate length bias and regular cells

regular cells (without any sleep transistors) has been divided by a factor of 10. As Fig. 5.8 shows, the new H/L cells that use gate-length biasing (instead of high-V_T devices) for the sleep transistors have a leakage that is between 1 and 2 orders of magnitude smaller than the leakage of regular cells. However, their leakage is between 1 and 2 orders of magnitude *greater* than the leakage of the regular H/L cells. We also simulated the new H/L cells and compared their delay impact. We found that the delay difference between the new H/L cells and the regular H/L cells was negligible. When compared with the regular H/L cells, the new H/L cells that used a gate-length biasing of 10% had between 1% and 3% smaller delay. For the new H/L cells that used a gate-length biasing of 20%, the delays were about the same as the regular H/L cells. Hence, we find that for the HL methodology, using high-V_T devices is more effective than using longer channel length devices since it gives a greater leakage reduction with a similar delay penalty. However, in case the cost associated with the additional threshold implant masks is to be avoided, one could use the H/L approach with gate-length biasing to obtain a leakage improvement over regular standard cells.

5.7 Leakage Reduction in Domino Logic

In this section we explore how leakage power reduction is achieved in dynamic cells. Specifically we focus on domino logic cells due to their widespread popularity.

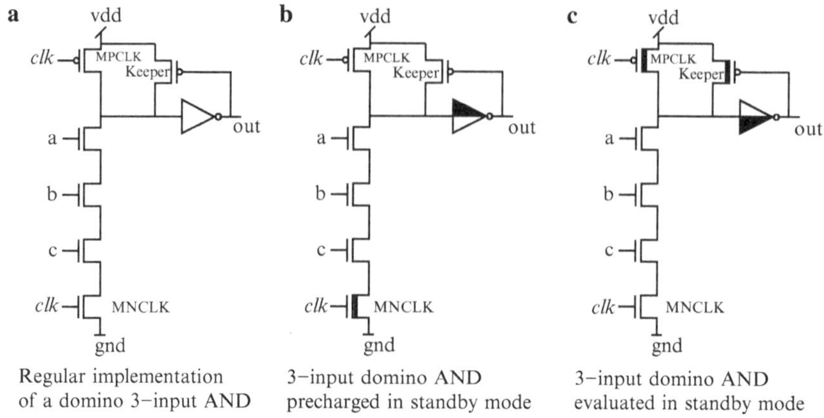

Fig. 5.9 Transistor level description (domino AND3 gate)

In standby mode, domino logic gates can either be in the *precharge* or *evaluate* state. In either case, if dual V_{T} technologies are used, devices that are turned *off* (devices in the cut-off mode of operation) in standby mode are implemented with high V_{T}. This can typically reduce leakage currents by about 2 orders of magnitude.

Figure 5.9 illustrates the low leakage alternatives for a domino logic AND3 gate. Figure 5.9a is a traditional domino AND3 gate. Figure 5.9b illustrates the design of an AND3 domino logic gate, which, in standby mode, is held in the precharge state (*clk* signal is logic-0). In this mode the PMOS pull-up device (MPCLK) is turned *on* and the NMOS pull-down stack is turned *off*. In the output inverter, the PMOS device is turned off and the NMOS device is turned on. The advantage in this method is that we have at least 2 devices turned *off* in series in the NMOS stacks, thus minimizing the leakage current. The footer device (MNCLK) and the PMOS device in the output inverter (illustrated by a dark triangle on the top part of the output inverter) are made high V_T to reduce leakage current further. However, both these devices are in the critical evaluate path of the domino logic gate, so the delay of the gate is increased when these devices are made high V_T. Therefore, these devices have to be up-sized to compensate for the increased delay. Rather than increasing the size of the footer device (MNCLK) alone, increasing the size of the rest of the devices in the NMOS stack results in smaller area penalties for the same delay.

Alternatively, the domino logic gate could be held in the *evaluate* state (NMOS stack turned *on*) during standby. In [8], the authors suggest such a method for a clock delayed domino logic scheme. An AND3 domino logic gate that is held in the *evaluate* state during standby is shown in Fig. 5.9(c). In standby mode the *clk* line is pulled high, thus turning off the PMOS pull-up device (MPCLK) and the NMOS in the output inverter. These devices are implemented with high V_{T} devices to keep the leakage current low. The *keeper* device is also made a high V_{T} device. The advantage of this scheme is that only the devices in the precharge path are made high V_{T} and

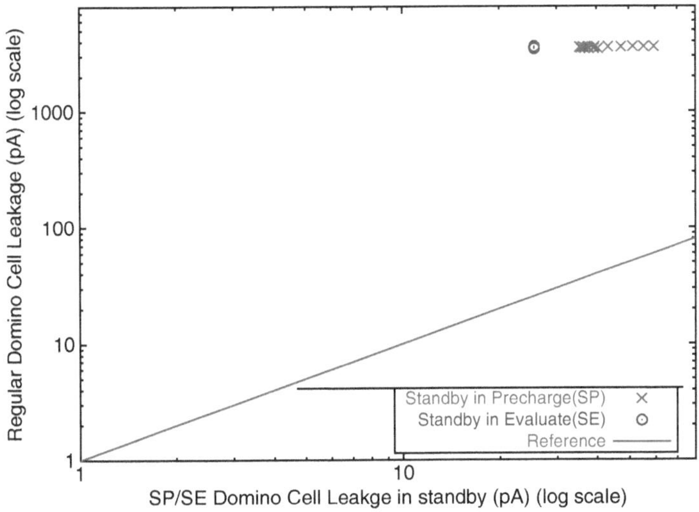

Fig. 5.10 Leakage of SE/SP versus regular domino cells

any delay increase is exhibited only in that path. We found that the delay in the evaluate mode is in fact decreased slightly due to reduced leakage contention from the high V_T PMOS device (MPCLK).

A comparison of the leakages of the different schemes for a library of cells (cells compared were AND2 AND3 AND4 AND5 AND6 AOI21 AOI22 OAI21 OAI22 OR2 OR3 OR4 OR5 OR6 OR7 and OR8) is shown in Fig. 5.10. The scheme in which cells are held in *precharge* during standby is referred to as SP, and SE denotes the scheme in which cells are held in *evaluate* state during standby. In a regular domino logic gate, all devices are low V_T devices. Devices in the evaluate path of SP gates were up-sized such that the gate delay (in the evaluation phase) was made equal to the regular domino logic gate. As can be seen from Fig. 5.10, the leakage of SP and SE cells is dramatically lower (by about 2 orders of magnitude) than that of regular domino logic cells (for the same delay). Also it can be seen that the leakage for the SE scheme does not change much across the different gates. This is because the leaking devices, the PMOS pull-up device (MPCLK) and the NMOS device in the output inverter, are of the same size for all gates. Leakage for SE cells was determined to be lower than for the SP cells, as illustrated in Fig. 5.10. This is because the high V_T devices in the SP cells had to be up-sized in order to avoid increased gate delays.

We also compared leakages of the SE and SP schemes for a set of circuits. The results are shown in Table 5.5. The leakage for the SE scheme is on average 31% lower than for the SP scheme.

From the above, it is clear that using SE domino logic gates is a better option from a delay, leakage and cell area standpoint (as compared to SP domino logic gates). For an SE domino logic gate, we need to ensure that all inputs of the gate are at logic-1 during standby mode. This can be done by gating the inputs of the

Table 5.5 Leakage comparison SE vs SP

Ckt.	SP Leakage(pA)	SE Leakage(pA)	Ovh (%)
alu2	17,516.82	12,290.93	−29.83
alu4	36,614.08	25,913.37	−29.23
apex6	21,543.77	15,261.24	−29.16
apex7	7,266.66	5,146.83	−29.17
dalu	82,461.74	58,253.88	−29.36
des	166,870.79	112,001.09	−32.88
C1355	29,497.98	21,099.43	−28.47
C1908	27,958.99	19,588.67	−29.94
C3540	60,278.10	41,968.40	−30.38
C432	9,184.09	6,401.53	−30.30
C499	23,250.06	16,592.75	−28.63
C6288	165,015.41	118,914.73	−27.94
C880	14,452.54	10,140.02	−29.84
i2	3,430.88	2,048.49	−40.29
i5	7,455.75	5,095.62	−31.66
i6	10,397.28	6,913.66	−33.51
i7	12,963.03	8,501.24	−34.42
i8	52,224.02	34,542.67	−33.86
i9	16,348.30	10,626.55	−35.00
i10	101,053.51	69,443.76	−31.28
t481	47,207.10	30,164.03	−36.10
too_large	17,053.89	11,650.78	−31.68
vda	19,747.06	12,777.45	−35.29
x3	23,492.55	16,157.45	−31.22
Avg			−31.64

first gate in a chain of domino logic cells. However, this will increase the delay of the gate during normal operation. The authors of [10] suggest a simple and elegant alternative. In this approach, an NMOS switch NS (as shown in Fig. 5.11) is used to pull down the dynamic node of the first gate in the chain. This switch is controlled by the standby signal. The only disadvantage of this method is that an additional standby signal is needed for the first gates in a chain of domino logic cells.

5.8 Summary

In this chapter, we have described low-leakage standard-cell-based ASIC design methodologies for both static CMOS and domino logic. The major contribution is the development of a new methodology for low-leakage static CMOS designs, which we call the "HL" methodology. This "HL" methodology is based on ensuring that during standby operation, the supply voltage is applied across more than one *off* device and there is at least one *off* device with a high V_T in the leakage path. For each standard cell in a library, we design two variants, the "H" and the "L" variant.

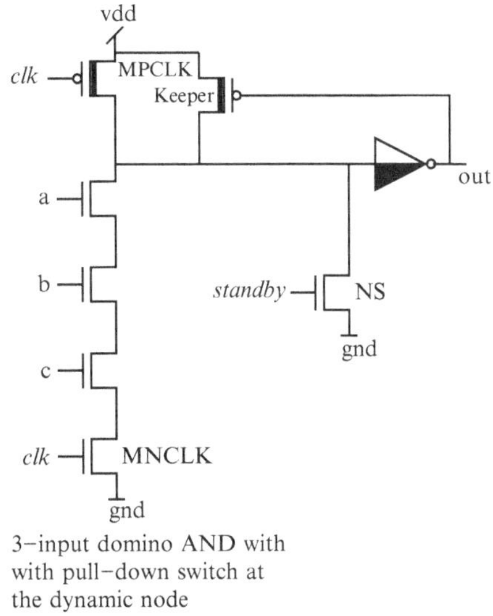

Fig. 5.11 Transistor level description of first SE domino gate in a chain

Our HL cells exhibit low-leakage currents as do MTCMOS gates, but with the advantage that leakage currents in our methodology can be precisely estimated (unlike MTCMOS). We compared the two techniques using 24 placed-and-routed designs. We have shown that our methodology has a lower delay than MTCMOS, which is expected since our HL cells exhibit a delay degradation for only one output transition. Our HL designs exhibit *predictable* leakage values that are much lower than the maximum leakage for MTCMOS designs. Since leakage in MTCMOS designs is not precisely controllable, this is a significant improvement. Further, our HL designs exhibit an area overhead of approximately 21–29% and 11–27% over regular designs (for delay-optimal and area-optimal mapping, respectively) and an area saving of up to 17% over MTCMOS designs. The HL methodology utilizes existing mapping and place/route tools and handles memory elements without additional routing overhead (unlike MTCMOS). We also explored the use of header and footer devices with long channel length instead of high-V_T devices in the H/L cells. We found that a higher V_T device was more effective. It gave a smaller leakage with a similar delay penalty.

With the downward scaling of VDD in future technologies, the threshold voltages of both the high V_T and low V_T devices in the HL methodology will have to scale down as well, if circuit delays are to be kept within reasonable limits. However, this could increase the leakage current. So if leakage current is the overriding concern, the V_T of the high-V_T power supply gating devices should not be scaled down. Though this may cause an increase in delays, this increase is in only one transition for each gate unlike traditional MTCMOS. Hence, the problems due to scaling of VDD in future technologies are similar for both MTCMOS and HL methodologies, but are worse for MTCMOS.

References

1. BSIM3 Homepage. http://www-device.eecs.berkeley.edu/~bsim3/intro.html. Accessed on 5th June 2004
2. Burd, T.: CMOS Standard Cell 2_3lp Library Documentation. University of California, Berkeley (1994)
3. Cadence Design Systems, Inc., 555 River Oaks Parkway, San Jose, CA 95134, USA: Envisia Silicon Ensemble Place-and-Route Reference (1999)
4. Cao, Y., Sato, T., Sylvester, D., Orshansky, M., Hu, C.: New Paradigm of Predictive MOSFET and Interconnect Modeling for Early Circuit Design. In: Proc. IEEE Custom Integrated Circuit Conference, pp. 201–204. Orlando, FL (2000). http://www-device.eecs.berkeley.edu/ ptm
5. Gupta, P., Kahng, A.B., Sharma, P., Sylvester, D.: Gate-Length Biasing for Runtime-Leakage Control. IEEE Transactions on Computer-Aided Design of Integrated Circuits and Systems **25**(8), 1475–1485 (2006)
6. Horiguchi, M., Sakata, T., Itoh, K.: Switched-Source-Impedance CMOS Circuit for Low Standby Subthreshold Current Giga-scale LSI's. IEEE Journal of Solid-State Circuits **28**(11), 1131–1135 (1993)
7. Kahng, A.B., Muddu, S., Sharma, P.: Impact of Gate-Length Biasing on Threshold-Voltage Selection. In: Proc. International Symposium on Quality Electronic Design, pp. 27–29. Santa Clara, CA (2006)
8. Kao, J.T., Chandrakasan, A.P.: Dual-Threshold Voltage Techniques for Low-Power Digital Circuits. IEEE Journal of Solid-State Circuits **35**(7), 1009–1018 (2000)
9. Kumagai, K., Iwaki, H., Yoshida, H., Suzuki, H., Yamada, T., Kurosawa, S.: A Novel Powering-down Scheme for Low Vt CMOS Circuits. In: Digest of Technical Papers, Symposium on VLSI Circuits, pp. 44–45. Honolulu, HI (1998)
10. Kursun, V., Friedman, E.G.: Low Swing Dual Threshold Voltage Domino Logic. In: Proc. IEEE Great Lakes Symposium on VLSI, pp. 47–52. New York, NY (2002)
11. McGeer, P.C., Saldanha, A., Brayton, R.K., Sangiovanni-Vincetelli, A.L.: Delay Models and Exact Timing Analysis, Chap. 8. Logic Synthesis and Optimization. Kluwer Academic Publishers, New York, NY (1993)
12. Min, K.S., Kawaguchi, H., Sakurai, T.: Zigzag Super Cut-off CMOS (ZSCCMOS) Block Activation with Self-adaptive Voltage Level Controller: An Alternative to Clock-Gating Scheme in Leakage Dominant Era. In: Digest of Technical Papers, International Solid-State Circuits Conference, vol. 1, pp. 400–502. San Francisco, CA (2003)
13. Mutoh, S., Douseki, T., Matsuya, Y., Aoki, T., Shigematsu, S., Yamada, J.: 1-V Power Supply High-Speed Digital Circuit Technology with Multithreshold-Voltage CMOS. IEEE Journal of Solid-State Circuits **30**(8), 847–854 (1995)
14. Nagel, L.: SPICE: A Computer Program to Simulate Computer Circuits. In: University of California, Berkeley UCB/ERL Memo M520 (1995)
15. Rabaey, J.: Digital Integrated Circuits: A Design Perspective. Prentice Hall Electronics and VLSI Series. Prentice Hall, Upper Saddle River, NJ (1996)
16. Sentovich, E.M., Singh, K.J., Lavagno, L., Moon, C., Murgai, R., Saldanha, A., Savoj, H., Stephan, P.R., Brayton, R.K., Sangiovanni-Vincentelli, A.L.: SIS: A System for Sequential Circuit Synthesis. Tech. Rep. UCB/ERL M92/41, erl, University of California, Berkeley, CA 94720 (1992)
17. Takashima, D., Watanabe, S., Nakano, H., Oowaki, Y., Ohuchi, K., Tango, H.: Standby/Active Mode Logic for Sub-1-V Operating ULSI Memory. IEEE Journal of Solid-State Circuits **29**(4), 441–447 (1994)

Chapter 6
Simultaneous Input Vector Control and Circuit Modification

6.1 Overview

Leakage power currently comprises a large fraction of the total power consumption of an IC. Techniques to minimize leakage have been researched widely. However, most approaches to reducing leakage have an associated performance penalty. In this chapter, we present an approach that minimizes leakage by simultaneously modifying the circuit while deriving the input vector that minimizes leakage. In our approach, we selectively modify a gate so that its output (in sleep mode) is in a state that helps minimize the leakage of other gates in its transitive fanout. Gate replacement is performed in a slack-aware manner, to minimize the resulting delay penalty. One of the major advantages of our technique is that we achieve a significant reduction in leakage without increasing the delay of the circuit.

The remainder of this chapter is organized as follows: The motivation for this work is described in Sect. 6.3. Section 6.4 discusses some previous work in this area. In Sect. 6.5, we describe our method to minimize leakage in a circuit through simultaneous input vector control and circuit modification. In Sect. 6.6, we present experimental results, while conclusions and future work are discussed in Sect. 6.7.

6.2 Introduction

One of the techniques used to minimize leakage is the technique of *parking* a circuit in its minimum leakage state. This technique involves very little or no circuit modification and does not require additional power supplies. A combinational circuit is *parked* in a particular state by driving the primary inputs of the circuit to a particular value during standby. This value can be scanned in via scab-enabled flip-flops or forced using MUXes (with the standby/sleep signal used as a select signal for the MUX). This technique for leakage reduction is frequently referred to as *input vector control*. In this chapter, we propose an approach that modifies and improves this technique to substantially achieve control over the leakage of a circuit at a finer granularity. We present an approach that minimizes leakage by *simultaneously* modifying the circuit while deriving the input vector that minimizes leakage.

N. Jayakumar et al., *Minimizing and Exploiting Leakage in VLSI Design*,
DOI 10.1007/978-1-4419-0950-3_6, © Springer Science+Business Media, LLC 2010

In our approach, we selectively modify a gate so that its output (in sleep mode) is in a state that helps minimize the leakage of other gates in its transitive fanout. Gate replacement is performed in a slack-aware manner, to minimize the resulting delay penalty. One of the major advantages of our technique is that we achieve a significant reduction in leakage without increasing the delay of the circuit. The leakage reduction technique discussed in this chapter is orthogonal to other circuit level leakage reduction approaches that statically (or dynamically) change the V_T of the devices.

The salient features of our technique are as follows:

- There are no floating nodes and hence no spurious transitions to deal with.
- Our technique does not require multiple V_T devices and this saves in mask costs.
- We extend input vector control to a finer granularity allowing us to control logic values at internal nodes.
- Our technique achieves significantly lower leakage with *zero delay penalty* unlike other techniques that offer similar leakage reduction at the expense of delay.
- Our algorithm is simple and involves a single linear pass of the circuit with additional static timing analysis runs over small sections of the circuit to check the timing slack.

6.3 The Intuition Behind Our Approach

Table 6.1 shows the leakage of a NAND3 gate for all possible input vectors to the gate. The leakage values shown are from a SPICE [10] simulation using the 0.1 μ BPTM [5] models at 1.2 V.

As can be seen from Table 6.1, setting a gate in its minimal leakage state (000 in the case of the NAND3 gate) can reduce leakage by about 2 orders of magnitude. This leakage reduction is attributed to the *stack effect*, according to which having as many *off* transistors in series as possible minimizes leakage. While it is desirable to set every gate in a circuit to its minimal leakage state, it may not be possible to do so due to the logical inter-dependencies of the inputs of the gates. Even if the individual gates have a wide range of leakage values, this does not mean that a multi-level circuit that uses these gates will have a wide range of leakage values

Table 6.1 Leakage of a NAND3 gate

Input	Leakage(A)
000	1.37e−10
001	2.70e−10
010	2.70e−10
011	4.96e−09
100	2.62e−10
101	2.68e−09
110	2.51e−09
111	1.01e−08

as well. For example, if a NAND3 gate and a NOR3 gate in a circuit share inputs, the leakage of the NAND3 is minimum when all the inputs are set to logic 0, but to get the NOR3 gate into its minimum leakage state requires all the inputs to be set to logic 1. Because of such constraints, we are limited in terms of the leakage reduction that we can achieve by using just vector control at the primary inputs. In order to exploit the stack effect better, we need a technique that offers more freedom in setting the inputs at each gate. Herein lies the key contribution of this chapter.

In practice, gate leakage currents can also contribute to the total leakage of a gate. However, the contribution of gate leakage only affects the table of leakage values for each input vector for a gate. Our algorithm is agnostic to this and only requires a reliable estimate of leakage currents of a gate for different input vectors, and hence it can account for gate leakage as well.

6.4 Related Previous Work

In an effort to exploit input vector control to minimize leakage, the problem of finding the minimum leakage sleep vector for a combinational CMOS gate-level circuit has received some attention recently. There are several heuristics ([3,4,7–9,11,14]) that have been proposed to find the minimum leakage sleep vector. Some of these have been discussed in Chap. 2 and also in Chaps. 3 and 4. While these heuristics attempt to find the minimum leakage vector assuming that only the primary inputs of a combinational circuit can be controlled, *we focus on circuit modifications as well, to ensure that we are not restricted to the primary inputs alone to control leakage*.

Traditionally, input vector control has involved using MUXes or scan chains to control the primary input values of a circuit during *standby*. We extend this idea further and give ourselves the freedom to set the inputs of individual gates in a circuit. We modify the circuit such that we are not restricted to controlling just the primary inputs, but can also control the internal nodes of a circuit. While the idea of adding control points is similar to what is expressed in [1,2], we allow a greater degree of freedom. In [1,2], the authors insert either AND or OR gates to set the logic value of a particular line during *standby*. We, on the other hand, allow one input going to two or more different gates to be split (using pass-gate MUXes), so that each fanout can be set to different values during standby. This provides significantly more opportunities to control internal nodes and minimize leakage. Also in [1,2], the authors use a SAT-based algorithm to find control points and to minimize leakage. The accuracy of the algorithm is dependent on the number of quantization levels of leakage values. However, with a higher number of quantization levels the runtime also increases. The algorithm we use has significantly lower complexity and involves a single linear-time traversal of the circuit. In [13], a technique is presented, which involves gate replacement. However, in [13] a gate G is replaced by a different gate G' to only reduce the leakage of gate G, *but not to control other internal circuit nodes*. The authors of [6] improve on the implementation of [13] in terms of both leakage improvement and runtime of the gate replacement algorithm.

Previous approaches to minimize leakage through vector control and gate replacement [1, 2, 6, 13] have an associated delay penalty to get a reasonable leakage reduction. The authors of [13] do not mention the exact delay penalty, but do state that their algorithms constrain the delay to within 5%. In [6] the authors improve on the leakage reduction achieved in [13] with an average delay penalty of 4.4%. In [1, 2], for sequential circuits, the authors claim that *up to* 70% leakage reduction is possible with a 15% delay penalty and *up to* 39% leakage improvement with a delay penalty of less than 2%. For combinational circuits, they achieve an average leakage improvement of 25% with a 5% delay penalty. In our approach, we get a significant leakage reduction (as shown in Sect. 6.6) with *no delay penalty*.

6.5 Our Approach

The algorithm we use in this chapter to minimize leakage and find control points in the circuit is designed to make sure that we do not ever get a negative slack. We have a built-in static timer that allows us to test if a gate violates timing.

One of the sources of our flexibility in controlling internal nodes of a circuit stems from the fact that we create several different variants of each gate in the library. While it may be argued that the creation of different variants of a cell can be time consuming and expensive, it should be noted that this step is done up-front and only once. An example of the different variants is shown in Fig. 6.1. In the *snglmx* type

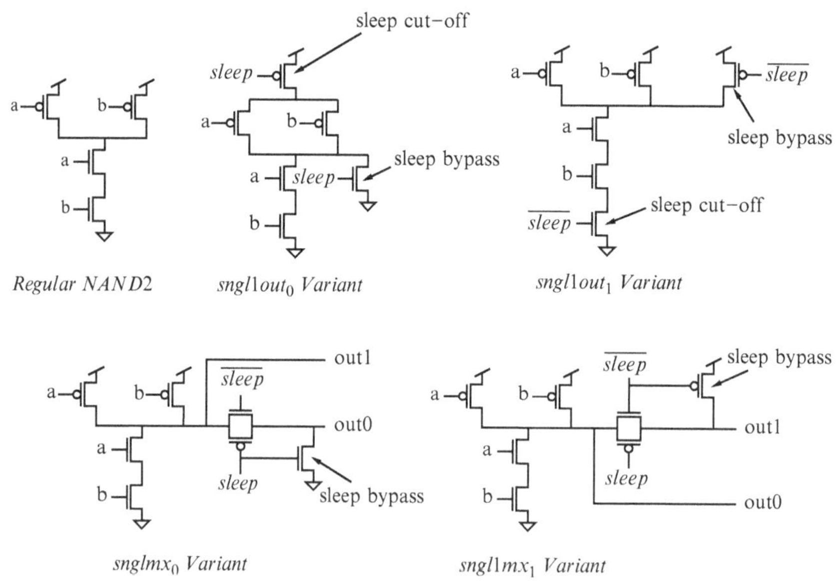

Fig. 6.1 Some variants of a NAND2 gate

of variant, a MUX is placed at the output of a regular gate. There are two types of *snglmx* gates, $snglmx_0$ and $snglmx_1$. The $snglmx_0$ gates have a weak pull-down device at the output of the MUX. A $snglmx_0$ variant is used as a replacement for a gate when the output of a gate G is logic 1 in standby, but some gates in the fanout of G require a logic 0 to get into a low-leakage state. Similarly, $snglmx_1$ gates have a weak pull-up device at the output of the MUX. A $snglmx_1$ variant is used when the output of a gate G is logic 0 in standby, but some gates in its fanout require a logic 1 to get into a low leakage state. Note that the *snglmx* type of variants are dual output gates and hence offer the most flexibility by "splitting" internal signals.

There can be situations when all the gates in the fanout of the gate in question need a value that is complementary to what is generated at the output of a gate in standby. For such cases we have a type of variant called the *sngl1out* variant. This type of variant has only one output and is similar to the structure discussed in [2]. We define two types of *sngl1out* variants, $sngl1out_0$ and $sngl1out_1$. The $sngl1out_0$ uses a PMOS sleep transistor to cut-off the PMOS stack of the gate (labeled as sleep cut-off in Fig. 6.1) and a weak NMOS pull-down device (labeled as sleep bypass in Fig. 6.1) to pull down the output. This variant is used when the output of a gate is high in the standby state, while all the gates in the fanout require a logic low value to get into a low-leakage state. Similarly the $sngl1out_1$ uses a NMOS sleep transistor to cut-off the NMOS stack of the gate and a weak PMOS pull-up device to pull up the output. This variant is used when the output of a gate is low in the standby state, while the gates in the fanout require a logic high value to get into a low-leakage state. Note that while the *snglmx* type of variant worsens both output rise and output fall delays, the *sngl1out* worsens delay for either only the rise or only the fall transition and can actually speed up the opposite transition. *In [2], the authors take advantage of this fact and assume the delay of such a gate to be the average of the rise and fall delays. This assumption can lead to inaccuracies in the timing analysis.* In our approach, we account for the rise and fall delays separately.

Because of the introduction of sleep devices, the delay of the *sngl1out* gates is larger than the regular cells (for one transition). Similarly, the *snglmx* variants also suffer a delay due to the pass gate MUX at the output. Since we have output timing constraints, this delay limits the flexibility of the gate replacement algorithm. To enhance the flexibility of the algorithm and give it more degrees of freedom, we also create larger cells that we call *dbl* cells. We create *dblmx* as well as *dbl1out* variants. Their structure and purpose is the same as their *sngl* counterparts except that they use larger device sizes ($\leq 2\times$ of their *sngl* counterparts). They are sized such that their delays are closer to the delays for regular gates. For the reader's reference, the area (active area), delay and leakage characteristics of the different variants are summarized in Tables 6.2, 6.3 and 6.4.

All these variants are crucial to our approach and help provide enough flexibility to our algorithm, reducing the leakage of a given circuit while making sure that there is *no delay penalty*. The details of the algorithm are explained in Sect. 6.5.1.

Table 6.2 Active Area (*in* μ^2) of some standard cells and their variants

Gate	Regular gate	$sngl1out_0$ variant	$sngl1out_1$ variant	$snglmx_0$ variant	$snglmx_1$ variant	$dbl1out_0$ variant	$dbl1out_1$ variant	$dblmx_0$ variant	$dblmx_1$ variant
INV1X	0.07	0.07	0.07	0.14	0.14	0.105	0.085	0.23	0.23
INV2X	0.14	0.14	0.14	0.23	0.23	0.21	0.175	0.42	0.42
NAND2	0.2	0.2	0.2	0.3	0.3	0.25	0.21	0.6	0.6
NAND3	0.36	0.36	0.36	0.46	0.46	0.525	0.42	0.96	0.96
NAND4	0.54	0.54	0.54	0.64	0.64	0.78	0.64	1.18	1.18
NOR2	0.27	0.27	0.27	0.37	0.37	0.36	0.3	0.65	0.65
NOR3	0.63	0.63	0.63	0.73	0.73	0.975	0.675	1.36	1.36

Table 6.3 Delay (*in ps*) assuming loading of five INV1X gates of some standard cells and their variants

Gate	Regular gate	$sngl1out_0$ variant	$sngl1out_1$ variant	$snglmx_0$ variant	$snglmx_1$ variant	$dbl1out_0$ variant	$dbl1out_1$ variant	$dblmx_0$ variant	$dblmx_1$ variant
INV1X	54.73	71.00	68.27	85.80	85.79	55.06	54.18	58.74	58.74
INV2X	33.18	46.90	41.28	58.74	58.74	34.23	34.01	42.29	42.29
NAND2	55.62	69.57	63.20	79.77	79.76	56.18	56.25	55.64	55.65
NAND3	63.81	81.72	73.65	93.05	93,04	63.25	64.32	66.78	66.78
NAND4	77.65	86.72	82.54	105.42	105.42	73.73	76.46	84.13	84.13
NOR2	64.58	80.56	73.13	94.14	94.13	67.21	65.23	70.65	70.65
NOR3	82.07	94.61	83.28	112.92	112.92	80.12	82.09	90.76	90.76

Table 6.4 Leakage characteristics (minimum : maximum) (*in nA*) of some standard cells and their variants

Gate	Regular gate	$sngl1out_0$ variant	$sngl1out_1$ variant	$snglmx_0$ variant	$snglmx_1$ variant	$dbl1out_0$ variant	$dbl1out_1$ variant	$dblmx_0$ variant	$dblmx_1$ variant
INV1X	1.7 : 2.8	0.4 : 4.6	0.2 : 1.7	1.7 : 2.8	1.7 : 2.8	0.5 : 4.8	0.2 : 1.8	3.3 : 5.6	3.3 : 5.6
INV2X	3.3 : 5.6	0.6 : 6.3	0.3 : 3.4	3.3 : 5.6	3.3 : 5.6	1.0 : 10.1	0.3 : 3.5	6.6 : 11.2	6.6 : 11.2
NAND2	0.2 : 6.7	0.6 : 5.8	0.1 : 4.7	0.2 : 6.7	0.2 : 6.7	0.7 : 7.7	0.2 : 6.1	0.4 : 13.5	0.4 : 13.5
NAND3	0.1 : 10.1	0.6 : 5.9	0.1 : 4.6	0.1 : 7.3	0.1 : 7.3	0.8 : 8.0	0.1 : 8.9	0.3 : 16.9	0.3 : 16.9
NAND4	0.1 : 13.5	0.9 : 9.8	0.1 : 10.8	0.1 : 7.3	0.1 : 7.3	1.0 : 10.1	0.1 : 10.9	0.2 : 7.4	0.2 : 7.4
NOR2	0.4 : 6.2	0.2 : 6.5	0.2 : 2.2	0.4 : 4.1	0.4 : 4.1	0.3 : 8.9	0.3 : 3.1	0.7 : 8.3	0.7 : 8.3
NOR3	0.3 : 10.1	0.2 : 14.0	0.6 : 6.5	0.3 : 7.5	0.3 : 7.5	0.4 : 19.7	0.8 : 7.8	0.6 : 7.8	0.6 : 7.8

6.5.1 The Gate Replacement Algorithm

Before we use the gate replacement algorithm, we first characterize our library of cells (including the variants) using SPICE [10], and generate a file in the GENLIB [12] format from the characterized data. In the GENLIB format, each pin of a gate is associated with an intrinsic delay component as well as a load-dependent component for both rise and fall times. Also included in the *genlib* file is the load capacitance of each input pin.

Algorithm replaceGateForMinLkg (levelized_netlist, genlib_data, allowed_slack)
find AT at all nodes
find RT at all nodes
set all gates at first level to minimum leakage state
for ($i = 1; i <= maxLevel\_of\_Ckt; i++$) **do**
  for ($j = 1; j <= num\_of\_gates\_at\_Level(i); j++$) **do**
    $G = G(j)$; pick a gate G from the gates at level i
    g = output signal of G
    find suggestedVal of g for all fanout(G)
    if all suggestedVal $= 0$ and logic value of g $= 1$ **then**
      $G_{new} = sngl1out_0$ variant of G
      CheckIfReplaceable(G, G_{new})
    else if all suggestedVal = 1 and logic value of g $= 0$ **then**
      $G_{new} = sngl1out_1$ variant of G
      CheckIfReplaceable(G, G_{new})
    else
      $G_{new} = snglmx_0$ variant of G
      CheckIfReplaceable(G, G_{new})
    end if
  end for
end for

Fig. 6.2 Algorithm to perform gate replacement

Algorithm CheckIfReplaceable (G,G_{new})
Check if G can be replaced by a $sngl$ variant
if G can be replaced by $sngl$ variant of G reduction in leakage and satisfying timing **then**
  replace G with the $sngl$ variant
else if G can be replaced by dbl variant of G with reduction in leakage and satisfying timing **then**
  replace G with the dbl variant
end if

Fig. 6.3 Algorithm to check to see if a gate is replaceable

The pseudo code for our algorithm is shown in Figs. 6.2 and 6.3. Our algorithm takes as input a netlist of gates in levelized order. We first perform a static timing analysis on this netlist to find the Arrival Times (ATs) and Required Times (RTs) at all nodes in the circuit. We use the cell characterization data (which accounts for the load dependency of both the rising and falling delays of the gates) for our static timing analysis. We assume that for gates driven by primary inputs, the primary input can be split to set the desired logic value at the inputs of these gates. Once the logic values of the inputs to the 0th level of gates (the gates with only primary inputs as the inputs) have been fixed, we propagate these values forward to the next level. Next, we pick a gate G from the 0th level. Lets say the output of the gate is a signal g. We then search through each of the gates h in the fanout of G and find the value of g that gives the minimum possible leakage for h. From this we get the logic value required of g for each h. For example, if one of these fanout gates is a two input gate H and assuming that one of its inputs is set to 1 due to another gate J, we would pick the minimum leakage from the following set of input vectors (11, 10). Thus, we get the value of g required to get this two input gate H in its minimum possible leakage state. Note that when we first visit any gate, we assume all possible input vectors are possible at each gate (i.e. we would consider all vectors 00, 01, 10 and 11 to get the minimum possible leakage vector). This step of finding the best value of g is done for all fanouts of G. If we need to set the value of g to 0 for some fanouts and to 1 in others (which would happen, for example, in situations where the signal g is an input to a NAND gate and a NOR gate), then we check if we can replace the gate G with its $snglmx$ variant. We first estimate the leakage savings (if any) of doing this replacement. The presence of the MUX and the weak pull-up/pull-down used in the $snglmx$ variant is a source of additional leakage. However, this increase could be outweighed by the leakage savings at the gates in the fanout of G. We estimate the difference and if there are savings, we then test if replacing G with a $snglmx$ variant causes timing violations. If there are timing violations, we attempt to use a $dblmx$ variant. Again we first check for leakage savings and if there are savings in leakage, we then check for timing violations. When checking for timing violations due to replacing G with a gate G', we first propagate new RTs at the gate G to its fanins. Also, note that replacing G implies changes in the capacitance seen by the gates in the fanin of G. We then recalculate the AT of the gates in the fanin of G. If the new AT is greater than the new RT, then we do not replace G with G'. If there is no timing violation (there is enough slack) and there are savings in leakage, then replace the gate G with its $dblmx$ variant. We follow a similar procedure if all the fanouts of G require the same value at g for minimum leakage. If this value required is the same as the value at g due to fixing the logic values at the inputs of G, then we do not need to replace the gate. If however, these value differ, then we attempt to first replace the gate with its $sngl1out$ variant. If such a replacement does not reduce leakage current, then we do not replace the gate G and move on to the next gate in the netlist. If such a replacement does not work due to timing slack violations, we then check if a $dbl1out$ variant of G would help without sacrificing power or timing. In this way we traverse the netlist in levelization order from primary inputs to primary outputs and replace gates as we move along, reducing leakage *while guaranteeing that there are no timing slack violations*. The complexity of the algorithm is $O(n^2)$, where n is the number of gates in the design. In some technologies, gate leakage can contribute to the total leakage. This would only change the leakage table look-up values and not affect the implementation of the algorithm.

6.6 Experimental Results

We performed extensive experiments to validate our method and compare its results to the minimum circuit leakage values. We simulated the circuits for 10,000 random vectors to find the minimum leakage (as suggested in [8]). Simulating 10,000 random vectors gives us over 99% confidence that less than 0.5% of the vector population has a leakage lower than the minimum leakage found through this random search. We assumed a library with the following basic cells: INV1X, INV2X, NAND2, NAND3, NAND4, NOR2, NOR3. The circuits for our simulations are from the ISCAS85 and MCNC91 benchmark suites. We first performed a technology-independent synthesis on these circuits in SIS [12] using *script.rugged* before mapping it with our library.

In Table 6.5, Column 2 and Column 3 show the minimum leakage current in nA for the original circuit and for the circuit modified by our algorithm, respectively. The % decrease in leakage current is shown in Column 4. The decrease in leakage current is 29.18% on average. *Note that this is the leakage decrease compared to the leakage obtained by applying input vector control alone.*

The critical delays (in ps) for the original and the modified circuit are shown in Columns 5 and 6, respectively. Column 7 gives the % decrease in critical delays of the modified circuit. We conjecture that one of the reasons for the delay decreasing is due to the fact that when the algorithm can not choose a *sngl* variant due to timing issues, it chooses a *dbl* variant and this can cause a decrease in the delay. Also, as mentioned in Sect. 6.5, while the delay of one type of transition gets worse in the *sngl1out* variants, the delay of the opposite transition is sped up slightly. The last Column of Table 6.5 reports the runtimes of the algorithm. The algorithm

Table 6.5 Leakage, delay improvements and runtimes for our approach

Ckt.	Original min lkg (nA)	New min lkg (nA)	% Lkg decr	Original delay	New delay	% Delay incr	Runtime (s)
alu2	1,251.72	1,022.44	−18.32	1,460.70	1,422.16	−2.64	5.53
alu4	2,598.14	2,094.99	−19.37	1,755.99	1,753.09	−0.17	21.16
apex6	2,743.08	1,753.82	−36.06	739.94	739.93	−0.00	20.03
apex7	812.72	592.88	−27.05	704.11	704.11	0.00	2.89
C1355	2,003.61	1,697.87	−15.26	930.41	930.23	−0.02	7.8
C432	584.46	449.93	−23.02	1,110.89	1,110.89	0.00	1.03
C880	1,375.73	977.07	−28.98	1,803.93	1,718.75	−4.72	6.12
C1908	1,909.95	1,548.12	−18.94	1,489.95	1,488.61	−0.09	10.1
C3540	4,079.92	3,126.00	−23.38	1,870.95	1,870.63	−0.02	51.89
C6288	13,020.10	12,011.39	−7.75	5,651.08	5,637.02	−0.25	695.85
dalu	3,293.89	2,378.24	−27.80	1,506.29	1,504.32	−0.13	42.75
des	15,218.02	12,013.16	−21.06	3,021.52	2,470.33	−18.24	655.38
i10	8,738.32	6,318.98	−27.69	2,549.68	2,499.43	−1.97	238.13
i1	158.38	102.96	−35.00	353.61	353.21	−0.11	0.11
i2	372.66	98.72	−73.51	392.98	392.98	0.00	0.51
i3	323.05	60.13	−81.39	182.46	182.46	0.00	0.98
i6	1,907.06	1,650.16	−13.47	1,080.10	1,080.10	0.00	5.5
i7	2,499.20	1,973.08	−21.05	1,088.31	1,088.31	0.00	10.38
i8	3,805.49	2,321.63	−38.99	1,591.76	1,297.01	−18.52	38.62
i9	2,552.20	1,440.26	−43.57	1,651.78	1,618.21	−2.03	15.87
t481	2,915.54	2,409.63	−17.35	901.69	838.36	−7.02	28.21
too_large	1,034.72	796.34	−23.04	680.24	677.89	−0.35	4.09
Avg			−29.18			−2.56	84.68

is currently implemented in PERL and was run on an Intel Pentium 4 with 2 GB of RAM, running Linux Fedora Core 3. The runtimes are expected to improve substantially when the algorithm is implemented in a compiled language such as C/C++.

Our algorithm assumes that there are MUXes at the primary inputs. They help ensure that all 0th level gates can be set independently into their low leakage state. For a fair comparison, we give the same flexibility (ability for the inputs of each of the 0th level gates to be set independently) when finding the minimum leakage vector for the original circuit.

In Table 6.6, the area penalty associated with using our algorithm is given. Note that this table refers to only the active area. Column 2 of the table shows the area of the original circuit. Column 3 and Column 4 of the table give the total area and the area overhead respectively of the modified circuit including the area of the sleep cut-off transistors used in the $sngl1out$ and the $dbl1out$ type of gates. The active area of these sleep cut-off transistors is reported in Column 5. Column 6 (which is obtained by subtracting Column 5 from Column 3) and Column 7 report the area and area overhead respectively of the modified circuit *excluding* the sleep cut-off transistors. On average, the total active area overhead including the sleep cut-off

Table 6.6 Area (active area) cost of using our approach

Ckt.	Original area (μ^2)	Total new area (μ^2)	Total new area Ovh (%)	Sleep transistor area (μ^2)	New area excluding sleep cut-off transistors (μ^2)	Area overhead excluding sleep cut-off transistors (%)
alu2	78.52	96.20	22.52	14.08	82.12	4.58
alu4	155.42	187.94	20.92	24.87	163.07	4.92
apex6	157.36	197.15	25.29	34.71	162.44	3.23
apex7	49.04	66.32	35.24	15.05	51.27	4.55
C1355	108.20	133.74	23.60	22.34	111.40	2.96
C432	37.92	46.01	21.33	7.29	38.72	2.11
C880	83.94	107.56	28.14	20.52	87.04	3.69
C1908	104.21	134.74	29.30	26.95	107.79	3.44
C3540	246.42	305.13	23.83	48.84	256.29	4.01
C6288	672.99	970.35	44.18	260.06	710.29	5.54
dalu	211.55	259.04	22.45	38.50	220.54	4.25
des	812.09	1054.80	29.89	209.27	845.53	4.12
i10	490.08	621.40	26.80	109.84	511.56	4.38
i1	11.90	13.99	17.56	1.85	12.14	2.02
i2	50.84	53.99	6.20	2.81	51.18	0.67
i3	32.28	40.36	25.03	5.00	35.36	9.54
i6	109.22	124.21	13.72	13.49	110.72	1.37
i7	147.63	170.96	15.80	21.11	149.85	1.50
i8	234.59	273.09	16.41	32.37	240.72	2.61
i9	151.56	179.53	18.45	24.13	155.40	2.53
t481	166.08	213.81	28.74	40.15	173.66	4.56
too_large	62.51	80.85	29.34	15.40	65.45	4.70
Avg			23.85			3.69

transistors is about 23.6%. However, the active area overhead excluding the sleep cut-off transistors is only about 3.7%, which implies that the sleep cut-off transistors caused most of the active area penalty. The size of the sleep transistors can be reduced by sharing them as is done in many MTCMOS-based designs. This would not only save area but also reduce leakage. Hence, we consider the active area *excluding* the sleep-cut off transistors (Columns 6 and 7 of Table 6.6) to be a more meaningful measure of the area penalty. Another important point to note is that the area overhead reported is only the active area overhead. The effective area overhead is expected to be much smaller once the circuits are placed and routed.

We also estimated the dynamic power consumption associated with using our approach. Intuitively, the dynamic power overhead is expected to be proportional to the active area overhead excluding the sleep transistors (3.7%). However, some of this active area is devoted to the sleep bypass transistors, which contribute only their diffusion capacitance to the total switched capacitance during circuit operation. Based on this we estimated the total switched capacitance overhead, which is proportional to the dynamic power consumption overhead. The switched capacitance overhead is shown in Column 8 of Table 6.7. The average switched capacitance overhead is

Table 6.7 Statistics of replacement gates utilized and switched capacitance overhead of using our approach

Ckt.	Number of sngl1out	Number of dbl1out	Number of snglmx	Number of dblmx	Total number of replacements	Total number of gates	Switched capacitance Ovh. (%)
alu2	91	0	30	0	106	374	2.42
alu4	183	2	66	0	218	713	2.68
apex6	204	0	18	0	213	779	1.06
apex7	94	0	6	0	97	255	1.38
C1355	91	16	0	0	107	582	1.38
C432	40	0	0	0	40	170	0.42
C880	119	0	12	0	125	404	1.31
C1908	150	3	6	0	156	548	1.04
C3540	327	0	58	0	356	1,174	1.69
C6288	1,649	2	70	0	1,686	3,578	1.53
dalu	342	0	36	0	360	946	1.53
des	1,171	0	170	0	1,256	4,169	1.64
i10	736	2	112	0	794	2,421	1.79
i1	12	0	0	0	12	52	0.40
i2	17	0	0	0	17	171	0.13
i3	4	60	0	0	64	114	6.37
i6	75	0	0	0	75	586	0.27
i7	111	0	0	0	111	719	0.30
i8	266	0	14	0	273	1,102	0.75
i9	167	2	4	0	171	735	0.73
t481	237	0	48	0	261	803	2.05
too_large	89	0	20	0	99	304	2.17
Avg	280.68	3.95	30.45	0.00	299.86	940.86	1.50

only about 1.5%, which is also roughly the dynamic power consumption penalty. Table 6.7 also shows statistics of the type (or variant) of the replacement gates used. We find that the *dblmx* variant of the gates did not get used at all. The *sngl1out* was the variant that was used the most. The next variant used most often was the *snglmx* variant. This variant along with the *dblmx* variant are the variants that offer the most flexibility in controlling the internal node voltages.

Tables 6.5, 6.6 and 6.7 validate the effectiveness of our methodology. Note, that the modified circuits have a lower leakage with *no delay penalty* (or in some cases a delay improvement) and a very small increase in dynamic power consumption. This is an improvement over previous approaches [1, 2, 6, 13] that obtain similar leakage improvements but at the expense of a delay increase. In [13], the authors claim an average leakage decrease of 17% for small circuits and 24% for large circuits. The area increase was 9% for small circuits and 7% for large circuits. The authors do not mention the exact delay penalty but restrict the delay penalty to less than 5% in their divide-and-conquer algorithm. In [6], the authors aim to improve on the approach in [13]. They achieve an average leakage reduction of 38% at the expense of an 18% area increase and a 4.4% delay penalty. In [1, 2] achieve an average leakage reduction of 25% with a delay penalty of 5% for combinational circuits. With a delay penalty of 15%, a higher energy savings of 45–50% is claimed with an area penalty of no more than 15%. For sequential circuits, the authors take advantage of existing scan chains to scan-in the lowest leakage vector, thus minimizing the area overhead. For sequential circuits they claim that *up to* 70% leakage reduction is possible with a 15% delay penalty and *up to* 39% leakage improvement is possible with a delay penalty of less than 2%. No area overheads are provided for the sequential circuits.

Our technique does not require multiple threshold voltages (which are required in MTCMOS-based methodologies) or multiple supply voltages (which are required in VTCMOS-based methodologies). Also, our technique does not suffer from the high currents drawn and the spurious transitions that occur when a MTCMOS circuit wakes up from the sleep mode. This is because in our technique, internal nodes do not float (outputs of gates are at full-rail values) when the circuit is put into the sleep state. In MTCMOS circuits, internal nodes float when the power gating sleep transistors are turned off.

We also performed experiments to test if our algorithm could reduce leakage even further if the allowed timing slack was increased. The results are shown in Table 6.8. We notice that not too many circuits (some exceptions are apex6, C432 and i9) are able to take advantage of the slack available. Our methodology currently only uses input vector control and circuit modification to allow control of internal node signals. However, if we allow the replacement of a gate with a lower leakage gate (through device sizing) or if we allow the the reduction of the size of the sleep cut-off transistors, then we could take advantage of the allowed slack. These features are not currently implemented since the primary goal was to decrease leakage with no delay penalty.

Table 6.8 Leakage improvement for different allowed slacks

	0% slack		10% slack		20% slack	
Ckt.	Lkg decr(%)	Delay incr(%)	Lkg decr(%)	Delay incr(%)	Lkg decr(%)	Delay incr(%)
alu2	−18.32	−2.64	−18.07	−2.28	−18.07	−2.28
alu4	−19.37	−0.16	−19.49	5.26	−19.49	5.26
apex6	−36.06	−0.00	−36.28	5.83	−36.21	18.34
apex7	−27.05	0.00	−28.39	6.87	−28.39	6.87
C1355	−15.26	−0.02	−24.08	4.73	−24.08	4.73
C432	−23.02	0.00	−33.13	9.22	−35.53	15.14
C880	−28.98	−4.72	−30.25	−6.45	−30.25	−6.45
C1908	−18.94	−0.09	−19.30	2.38	−19.30	2.38
C3540	−23.38	−0.02	−23.22	5.75	−23.22	5.75
C6288	−7.75	−0.25	−7.54	1.53	−7.54	1.53
dalu	−27.80	−0.13	−27.33	3.32	−27.33	3.32
des	−21.06	−18.24	−21.06	−18.24	−21.06	−18.24
i10	−27.69	−1.97	−27.69	−1.68	−27.69	−1.68
i1	−35.00	−0.11	−41.13	5.54	−41.13	5.54
i2	−73.51	0.00	−76.18	3.70	−76.18	3.70
i3	−81.39	0.00	−90.37	5.86	−90.37	5.86
i6	−13.47	0.00	−25.28	−7.91	−25.28	−7.91
i7	−21.05	0.00	−27.28	−5.61	−27.28	−5.61
i8	−38.99	−18.52	−38.91	−18.52	−38.91	−18.52
i9	−43.57	−2.03	−43.93	7.08	−44.00	11.56
t481	−17.35	−7.02	−17.35	−7.02	−17.35	−7.02
too_large	−23.04	−0.34	−24.11	6.04	−24.11	6.04
Avg	−29.18	−2.56	−30.28	0.25	−30.28	1.29

6.7 Summary

In this chapter we presented an algorithm that replaces gates in a circuit, in an effort to reduce the standby leakage of the circuit. This replacement does not necessarily reduce the leakage of a gate being replaced, but helps set the gates in the transitive fanout to their low-leakage states. The algorithm involves traversing the circuit from the primary inputs to the primary outputs, replacing gates as required to try and set as many gates as possible to their low-leakage state. We get an average decrease in leakage of about 29% with an active area penalty of about 24%. This leakage decrease is the decrease over the leakage obtained through input vector control alone.

Possible extensions to this work could be using a larger library with complex gates and implementing a "smarter" algorithm that starts with a solution (given an initial minimum leakage vector) and then replaces gates if required. This could potentially yield much lower leakage currents.

References

1. Abdollahi, A., Fallah, F., Massoud, P.: Runtime Mechanisms for Leakage Current Reduction in CMOS VLSI Circuits. In: Proc. 2002 International Symposium on Low Power Electronics and Design, pp. 213–218. Monterey, CA (2002)
2. Abdollahi, A., Fallah, F., Pedram, M.: Leakage Current Reduction in CMOS VLSI Circuits by Input Vector Control. IEEE Transactions on VLSI Systems **12**(2), 140–154 (2004)
3. Aloul, F., Hassoun, S., Sakallah, K., Blauuw, D.: Robust SAT-Based Search Algorithm for Leakage Power Reduction. In: Proc. Power and Timing Models and Simulation. Seville, Spain (2002)
4. Bahar, R.I., Frohm, E.A., Gaona, C.M., Hachtel, G.D., Macii, E., Pardo, A., Somenzi, F.: Algebraic Decision Diagrams and Their Applications. Formal Methods in Systems Design **10**(2/3), 171–206 (1997)
5. Cao, Y., Sato, T., Sylvester, D., Orshansky, M., Hu, C.: New Paradigm of Predictive MOSFET and Interconnect Modeling for Early Circuit Design. In: Proc. IEEE Custom Integrated Circuit Conference, pp. 201–204. Orlando, FL (2000). http://www-device.eecs.berkeley.edu/ ptm
6. Cheng, L., Deng, L., Chen, D., Wong, M.D.F.: A Fast Simultaneous Input Vector Generation and Gate Replacement Algorithm for Leakage Power Reduction. In: Proc. Design Automation Conference, pp. 117–120. San Francisco, CA (2006)
7. Gao, F., Hayes, J.: Exact and Heuristic Approaches to Input Vector Control for Leakage Power Reduction. In: Proc. International Conference on Computer-Aided Design, pp. 527–532. San Jose, CA (2004)
8. Halter, J., Najm, F.: A Gate-Level Leakage Power Reduction Method for Ultra Low Power CMOS Circuits. In: Proc. Custom Integrated Circuits Conference, pp. 475–478. Santa Clara, CA (1997)
9. Johnson, M., Somasekhar, D., Roy, K.: Models and Algorithms for Bounds on Leakage in CMOS Circuits. IEEE Transactions on Computer-Aided Design of Integrated Circuits and Systems **18**(6), 714–725 (1999)
10. Nagel, L.: SPICE: A Computer Program to Simulate Computer Circuits. In: University of California, Berkeley UCB/ERL Memo M520 (1995)
11. Rao, R., Liu, F., Burns, J., Brown, R.: A Heuristic to Determine Low Leakage Sleep State Vectors for CMOS Combinational Circuits. In: Proc. International Conference on Computer-Aided Design, pp. 689–692. San Jose, CA (2003)
12. Sentovich, E.M., Singh, K.J., Lavagno, L., Moon, C., Murgai, R., Saldanha, A., Savoj, H., Stephan, P.R., Brayton, R.K., Sangiovanni-Vincentelli, A.L.: SIS: A System for Sequential Circuit Synthesis. Tech. Rep. UCB/ERL M92/41, erl, University of California, Berkeley, CA 94720 (1992)
13. Yuan, L., Qu, G.: Enhanced Leakage Reduction Technique by Gate Replacement. In: Proc. Design Automation Conference, pp. 47–50 (2005)
14. Zhanping, C., Johnson, M., Liqiong, W., Roy, W.: Estimation of Standby Leakage Power in CMOS Circuit Considering Accurate Modeling of Transistor Stacks. In: Proc. International Symposium on Low Power Electronics and Design, pp. 239–244. Monterey, CA (1998)

Chapter 7
Optimum Reverse Body Biasing for Leakage Minimization

7.1 Overview

One of the methods to reduce leakage power is by increasing the threshold voltages (V_T) of the device. This is done either statically, through use of multi-threshold devices or dynamically, through Reverse Body Biasing (RBB).

The sub-threshold leakage (cut-off) current of a transistor decreases with greater applied RBB. Reverse Body Biasing affects V_T through body effect, and sub-threshold leakage has an exponential dependence on V_T, as we have discussed earlier.

However, while the sub-threshold leakage decreases, there are other components to the leakage current that have to be considered as well. Two of these are bulk Band-to-Band-Tunneling (BTBT) and surface BTBT. Bulk BTBT is commonly referred to as simply BTBT while surface BTBT is commonly called Gate Induced Drain Leakage (GIDL) [2, 8]. While GIDL does not play a major role at RBB [2], BTBT increases with applied RBB [2, 5, 6, 9]. This means that there is an optimum RBB voltage at which the *total* leakage power (the sum of the sub-threshold leakage, the gate leakage, BTBT and GIDL) is minimum [2,5,6,9]. In modern processes this optimum point is reached before the upper limit of the RBB (based on the voltage at which the bulk–drain/bulk–source junction breaks down). Also, this optimum point can vary with temperature and process variations. In this chapter we show that it is desirable to operate at the optimal RBB point that minimizes total leakage. We present a scheme that monitors the total leakage current (the sum of the sub-threshold, BTBT and gate leakage) of an IC with a representative leaking device and, using this monitored value, we *automatically* find the optimum RBB value across temperature and process corners, using a self-adjusting circuit. Our approach has a modest placed-and-routed area utilization and a low power consumption. In Sect. 7.2 we discuss the motivation behind our work. Section 7.3 discusses previous approaches to dynamically adjust body bias. Section 7.4 describes our approach to dynamically self-adjust the RBB of PMOS and NMOS devices in order to obtain a minimum total leakage, along with experimental results that support the utility of our scheme.

N. Jayakumar et al., *Minimizing and Exploiting Leakage in VLSI Design*,

DOI 10.1007/978-1-4419-0950-3_7,

7.2 Goal and Background

In this work we are concerned with minimizing the total leakage current (the sum of the sub-threshold, BTBT and gate leakage) through a non-conducting (turned-off) device in a static CMOS design. In the case of an NMOS device this would mean that we are concerned with minimizing the leakage (over possible RBB values) through an NMOS device when its drain terminal is at VDD, its source and gate terminals are at GND and its bulk terminal (p- well) is at a certain RBB value. In such a scenario, the leakage current measured at the drain of the device is mainly due to three sources – (1) the sub-threshold leakage from the drain to the source of the device, (2) the gate leakage current from the drain to the gate and (3) the drain–bulk junction current. The drain-bulk leakage current has three main components – bulk BTBT (or simply BTBT), surface BTBT (or GIDL) and the classical reverse-biased PN junction current [2, 7, 10] (see Fig. 1.2 in Chap. 1). The bulk BTBT current is also often referred to as Gate Edge Drain Leakage (GEDL). This current is due to the tunneling of electrons from the valence band of the p-region (from the bulk) to the conduction band of the n-region (to the drain). This tunneling happens due to a high electric field across the bulk–drain junction [which can happen when a Reverse Body Bias (RBB) is applied]. Gate Induced Drain Leakage current (GIDL) occurs when the gate bias is negative relative to the drain [1, 10]. At negative gate bias, the overlap region of the gate and drain gets depleted of carriers. Minority carriers (generated by BTBT and other tunneling mechanisms) arrive at the surface to attempt to form an inversion layer in the channel and are immediately swept laterally to the substrate. Because of the field across the gate and bulk junction, these carriers then flow into the bulk node. This current is the GIDL current.

The two BTBT currents dominate the reverse-biased PN junction current. While the sub-threshold leakage decreases with increased RBB (due to the increase in V_T of the device), bulk BTBT current increases with RBB. The BTBT current density equation [12] is given below

$$J_{\text{BTBT}} = A\frac{E\,V_{\text{app}}}{\sqrt{E_g}}e^{-B\left(\frac{E_g^{3/2}}{E}\right)}. \tag{7.1}$$

$$A = \frac{\sqrt{2m^*q^3}}{4\Pi^3\hbar^2}. \tag{7.2}$$

$$B = \frac{4\sqrt{2m^*}}{3q\hbar}. \tag{7.3}$$

In these equations, m^* is the effective mass of an electron, E_g is the energy band-gap, V_{app} is the applied reverse bias, E is the electric field at the junction, q is the electron charge, and $\hbar = 1/(2\Pi)$ times Planck's constant.

Assuming a step function, the electric field at the junction is

$$E = \sqrt{\frac{2qN_aN_d(V_{\text{app}} + V_{\text{bi}})}{\varepsilon_{\text{Si}}(N_a + N_d)}}, \tag{7.4}$$

where N_a and N_d are the doping in the P and N devices, ε_{Si} is the permittivity of silicon and V_{bi} is the built-in voltage across the junction. Hence for a step junction, J_{BTBT} is approximately proportional to $V_{app}^{3/2}$. However, the exact dependence of E on V_{app} varies with the doping profile of the substrate [9].

The drain–gate leakage current does not change appreciably with applied RBB [9]. Also, at RBB, bulk BTBT dominates GIDL [2]. Hence, it is mainly the sub-threshold and the BTBT component of the leakage currents that change with applied RBB. Also, since these two components behave differently with respect to RBB, there exists an optimal RBB value [2,6,9,10] that minimizes leakage. We performed experiments on a test chip manufactured using the TSMC 0.13 μm triple well process to find the RBB value that minimizes total leakage. The test chip had one large PMOS ($W_{eff} = 676$ mm, $L_{eff} = 0.13$ μm) and one large NMOS ($W_{eff} = 504$ mm, $L_{eff} = 0.13$ μm) device. The devices on the test chip were made large so that their different leakage current components would be easy to measure. The drain, source, gate and bulk contacts were all brought out as pins, enabling us to measure the currents at each of these contacts. When a device is turned-off, the current measured at the source represents the sub-threshold leakage current from the drain to the source (I_{ds}), the current measured at the gate represents the gate leakage from the drain to the gate (I_{dg}) and the current measured at the bulk contact represents the drain/source to bulk current (I_{db},I_{sb}). Since the drain is at VDD, most of the bulk current is from the drain (i.e. I_{db} dominates I_{sb}). The current measured at the drain of the device (I_{leak}) was found to be approximately the sum of the currents measured at the gate, source and bulk terminals confirming that I_{sb} is very small in practice.

Figure 7.1 shows measurements taken from our manufactured test chip for a non-conducting NMOS device at a temperature of 25°C with the RBB being swept from 0.7 to 1.1 V below the source terminal. The VDD used was 1.2 V. In this case the optimal RBB value is 1.0 V.

The optimum RBB value can shift with temperature and process variations. Table 7.1 shows the penalty due to temperature variations (in terms of percentage of leakage power increase from optimum) for the large NMOS device, while Table 7.2 reports the penalty due to process variations, assuming that the RBB is fixed to the optimum value (1.015 V) for one particular temperature and process corner (25°C and nominal corner in this case).

Tables 7.1 and 7.2 prove that fixing the RBB at a particular value may not be a good idea if we are interested in reducing leakage over *all* temperature and process variations. We hence need a scheme by which we can monitor the leakage current of a chip and automatically self-adjust the RBB value of the PMOS and NMOS devices, to keep the leakage power as low as possible. The problem of monitoring the optimum point is compounded by the fact that the total leakage current can vary by as much as 3 orders of magnitude over temperature and RBB variations. The leakage monitor must therefore be able to find the optimum RBB point over this wide range of currents.

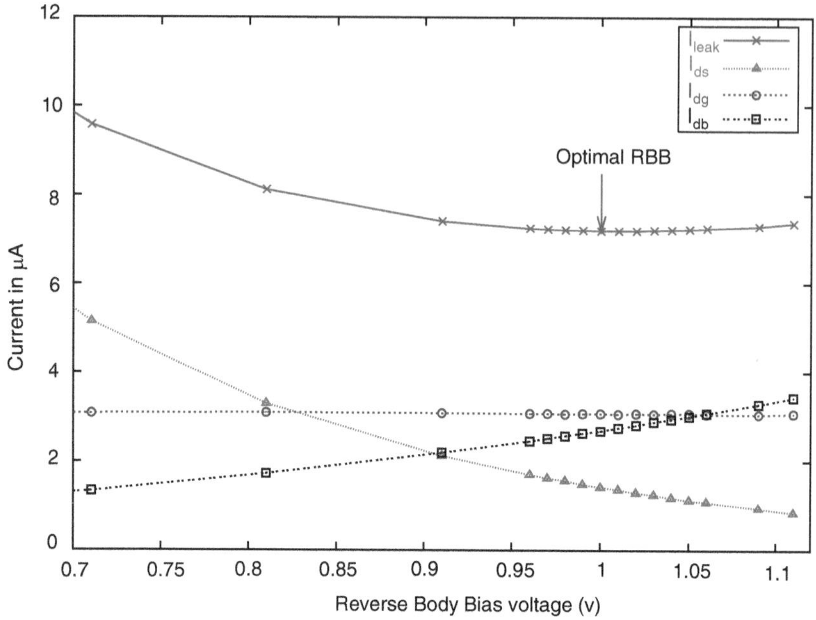

Fig. 7.1 Leakage current components for a large NMOS device at 25°C

Table 7.1 Leakage penalty due to temperature variation

Temp (°C)	Lkg penalty
−40	23.38%
0	6.99%
25	0%
70	35.29%
125	163.55%

Table 7.2 Leakage penalty due to process (V_T, l_{eff}) variation

V_T	l_{eff}	Lkg penalty (%)
Nominal	Nominal+10 nm	16.15
Nominal	Nominal−10 nm	4.02
Nominal	Nominal	0
Nominal−8%	Nominal−10nm	10.73
Nominal+8%	Nominal+10nm	58.3
Nominal+8%	Nominal	20.77

7.3 Related Previous Work

In [9], a simple circuit is presented that helps find the optimal RBB value. The accuracy of this circuit is dependent on the assumption that gate leakage can be neglected (or is very small) and that sub-threshold leakage is negligible when compared to the BTBT current in a stack of two non-conducting devices. Under these assumptions,

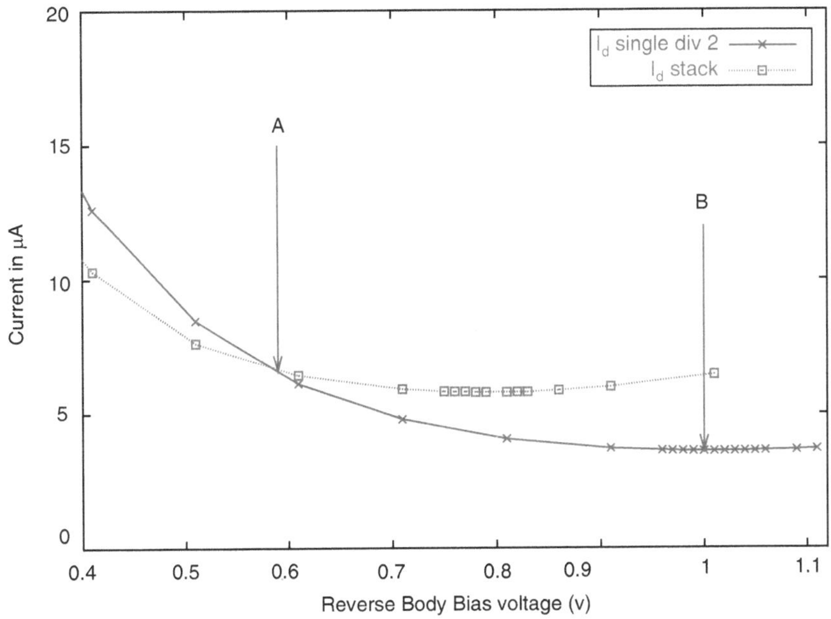

Fig. 7.2 Leakage current for stacked and single devices

the authors claim that the optimal RBB value occurs at the point where the leakage current through two stacked non-conducting devices is primarily BTBT current and is equal to half the leakage through a single non-conducting device. However, experiments with our test chip show that these assumptions are significantly inaccurate.

Figure 7.2 shows a plot of *half the leakage current* through a single non-conducting NMOS device on our test chip (labeled as "Id single div 2") and the leakage current through a stack of two non-conducting NMOS devices (labeled as "Id stack"). The currents were measured at a temperature of 25°C. The arrow labeled "A" shows the optimal RBB value as would be suggested by the circuit in [9] while the arrow labeled "B" shows the actual optimal RBB value for a single non-conducting NMOS device at 25°C. We found that if the RBB value marked by A was used as the "optimal" RBB instead of the RBB value pointed by B, the leakage current for a single non-conducting NMOS device (at 25°C) would be 70% higher than optimum.

In [4] and [3] the authors suggest sensing the voltage dropped by a leaking device towards the goal of adjusting the body bias and thus controlling the leakage. To amplify the leakage current, the gate bias is set to a value such that the leaking device is still cut-off but has a high enough leakage current to drop a significant voltage. This voltage is sensed and if it crosses a certain threshold, RBB is applied. The authors of [11] suggest a similar mechanism as a way of stabilizing sub-threshold CMOS logic. However, [3, 4, 11] do not target the problem of finding the *optimum* RBB value.

7.4 Leakage Monitoring/Self-Adjusting Scheme

Our leakage monitoring scheme is based on measuring the time taken for the leakage current to discharge (for monitoring the leakage of a leaking NMOS device) a capacitive load. For a leaking PMOS device, the time taken for charging-up the load is considered. A higher leakage would be indicated by a shorter time to discharge the load while a longer time to discharge the load would indicate a lower leakage. To monitor the leakage current of an NMOS device, the capacitively loaded node is initially precharged to a logic-high value. The leakage current is estimated by measuring the time taken to discharge this node. Similarly, for a leaking PMOS device, the capacitively loaded node is initially pre-discharged and the leakage current is estimated based on the time taken to charge this node to a logic-high value.

The leakage monitoring scheme is conceptually illustrated in Fig. 7.3 (for NMOS bulk control). A similar structure is used to control the PMOS bulk node. The three main blocks of the leakage monitoring scheme are (1) a leakage current monitoring (LCM) block that contains a representative leaking device, (2) a digital block to interface with the LCM and control the body bias voltage and (3) a programmable body bias voltage generator to translate the body bias control value from the digital block into a body bias voltage value. In this chapter we deal with the leakage monitoring block and the digital control block. Details of the bias generator are omitted, ant it is assumed that this function is performed by an off-the-shelf Digital to Analog converter (DAC) IC.

7.4.1 Leakage Current Monitoring Block (LCM)

In this section the design and operation of the LCM block will be discussed. We use the LCM for NMOS devices as an example. Our objective is to track the variation of

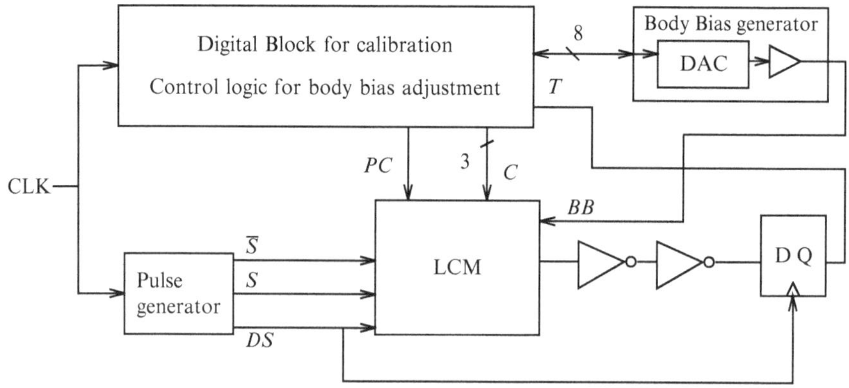

Fig. 7.3 LCM scheme block diagram (for NMOS)

total leakage current through a circuit with applied RBB. However, placing a current monitoring device in series with the IC supply and circuit power rails of the logic devices is not an option since the addition of such a device would increase the delay of the circuit. Hence, we choose a representative device to model the leakage of the entire circuit. The optimal RBB value is smaller for stacked devices when compared to single (unstacked) devices. This is because sub-threshold leakage is lower for stacked devices and hence BTBT dominates at a lower RBB value. However, it is infeasible to have separate substrates for stacked and non-stacked devices. In our scheme we chose a non-stacked device as the representative leaking transistor based on the intuition that for most ICs the dominant source of leakage is from unstacked devices. However, if we were to design a leakage monitor to track the leakage of an IC (with stacked devices being the dominant source of leakage), the leakage monitor would have to use stacked devices as the representative leaking transistors.

The leakage current variation of NMOS and PMOS devices is monitored separately. Figure 7.4 shows the circuit that implements the leakage current monitoring block for NMOS devices. In Fig. 7.4, device M_L is the representative leaking transistor. Transistor M_{pchg} is the device that precharges the node N_{chk}. M_L and M_{pchg}

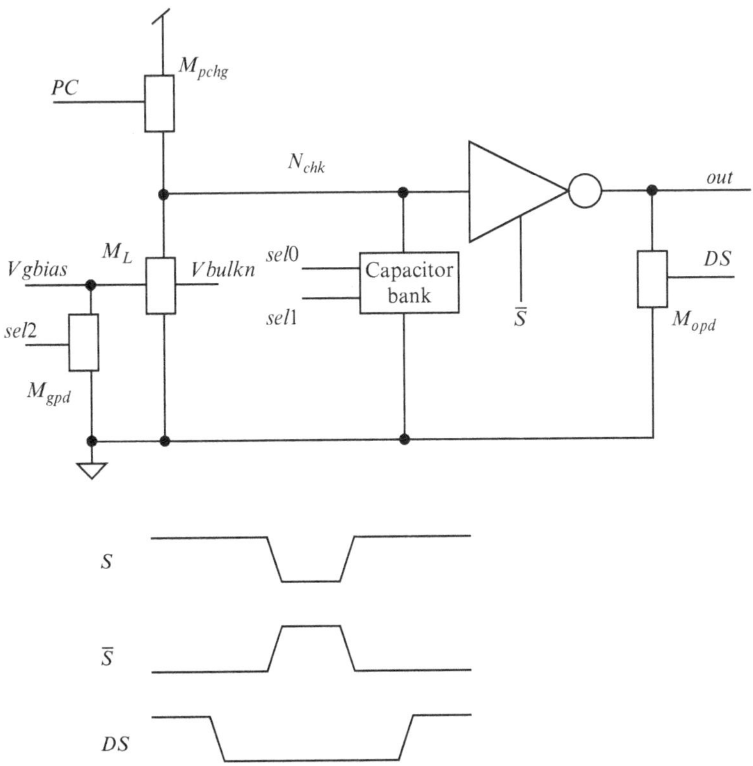

Fig. 7.4 LCM for NMOS devices

are sized relatively so that the leakage of M_L dominates the leakage of M_{pchg}. The leakage monitoring scheme is based on the idea that the time taken for the leaking transistor M_L to discharge the node N_{chk} would be proportional to the leakage current through M_L and hence the leakage current through the entire circuit.

In Fig. 7.4, the capacitor bank and the device M_{gpd} allow the LCM to work over a wide range of leakage currents. If the leakage current is too low, it needs to be magnified for the LCM to work effectively. This is done by first disconnecting the capacitor bank from N_{chk} (to speed up the rate of discharge of the node N_{chk}). Further magnification of the leakage current is achieved by turning off M_{gpd} and hence increasing the gate bias of M_L (in a similar manner as in [3, 4]) to a value of about 0.1 V above GND (such that M_L is still in the sub-threshold/cut-off mode).

The circuit that generates this low gate bias voltage is designed such that its output voltage decreases with an increase in temperature. Without this feature, the current in M_L increases too rapidly with increasing temperature when M_{gpd} is off.

The LCM works by "sampling" (turning on the tri-stateable inverter at the output of the LCM) the node N_{chk} at regular intervals. During this sampling, the output pull-down device, M_{opd} is turned off. Note that the sampling period is short, which keeps the power consumption of the LCM low. If the node N_{chk} has fallen low enough, the output of the LCM goes high and this output is buffered and then latched in a D flip-flop. The DFF output (shown as T in Fig. 7.3) triggers the digital block. The purpose of this trigger signal will be explained in the following sub-section.

The LCM for PMOS devices is implemented in a manner similar to that of the LCM for NMOS devices.

7.4.2 Digital Control Block

The Digital Control Block contains an 8-bit counter that counts up till either the end of the count is reached or till it receives a trigger signal from the DFF at the output of the LCM. When a trigger signal is received, the value of the 8-bit counter is stored. This counter value is proportional to the time taken for the transistor M_L to discharge the node N_{chk} and is hence a measure of the leakage current of M_L. Next, the node N_{chk} is precharged (signal PC goes low) and held in this precharged state till a new body bias is set. The applied RBB value is increased till the point at which the new counter value is smaller than the previous counter value (the point at which the leakage current starts increasing with applied RBB). If the end of the count is reached before a trigger signal is received, this implies that the total leakage is too low. In such a situation, control signals from the digital block are applied to the LCM to magnify the leakage current. The digital block sends appropriate signals (shown as C in Fig. 7.3 and $sel0$, $sel1$, $sel2$ in Fig. 7.4) that control the capacitor bank and M_{gpd} in the LCM to achieve this magnification, as described in Sect. 7.4.1.

In summary, our leakage monitoring scheme works by essentially converting the problem of sensing the total leakage current into one of measuring the time taken for a representative leaking transistor to discharge a purely capacitive load. The

Table 7.3 Size of the standard-cell implementations of the LCMs and pulse generator

Cell	Width (μm)	Height (μm)	Area (μm^2)
LCM NMOS	77.87	3.285	255.7
LCM PMOS	86.41	3.285	283.86
Pulse generator	38.22	3.285	125.55
Total	–	–	665.11

time taken is measured using a counter and the applied RBB is increased in linear steps till the time measured by the counter for a particular body-bias value is shorter than the time measured by the counter for a previous body-bias value used. The LCM is designed for correct operation over a wide range of leakage currents.

The accuracy of the scheme can be improved by increasing the frequency of the clock and hence increasing the frequency of sampling of the node N_{chk}. We utilize a clock with a period of 2 ns. Simulations showed the proposed scheme has a very small power consumption of 11.4 μA. Of this, the LCM block consumes about 4 μA, while the digital control block consumes about 6 μA. Note that simulations were done at 1.2 V at 125°C (to model the worst-case power consumption) for a TSMC 0.13 μm process. The digital block was synthesized using a 0.13 μm process standard-cell library.

We also created layout macro-cells for the pulse generator (that generates the S and DS signals for the LCM block), the LCM block for NMOS leakage monitoring and the LCM block for PMOS leakage monitoring. The LCM blocks include the circuitry required to generate the low *Vg bias* voltage. Table 7.3 shows the placed-and-routed size of each cell in the layout.

7.5 Summary

In this chapter, we have described an automatic, self-adjusting mechanism to find the optimal RBB value to minimize total leakage. Our method consists of a leakage current monitor and a digital block that senses the discharging (charging in the case of a PMOS transistor) of a representative NMOS device in the design. Based on the speed of discharge, which is faster for leakier devices, an appropriate RBB value is applied. Our technique is able to find the optimal RBB point and incurs very reasonable placed-and-routed area and power penalties in its operation.

References

1. Chen, J., Wong, S., Wang, Y.: An Analytic Three-Terminal Band-to-Band Tunneling Model on GIDL in MOSFET. IEEE Transactions on Electron Devices **48**(7), 1400–1405 (2001)

2. Keshavarzi, A., Narendra, S., Borkar, S., Hawkins, C., Royi, K., De, V.: Technology Scaling Behavior of Optimum Reverse Body Bias for Standby Leakage Power Reduction in CMOS ICs. In: Proc. International Symposium on Low Power Electronics and Design, pp. 252–254. San Diego, CA (1999)
3. Kobayashi, T., Sakurai, T.: Self-adjusting Threshold-Voltage Scheme (SATS) for Low-Voltage High-Speed Operation. In: Proc. IEEE Custom Integrated Circuits Conference, pp. 271–274. San Diego, CA (1994)
4. Kuroda, T., Fujita, T., Mita, S., Nagamatsu, T., Yoshioka, S., Suzuki, K., Sano, F., Norishima, M., Murota, M., Kako, M., Kakumu, M.K.M., Sakurai, T.: A 0.9-V, 150-MHz, 10-mW, 4 *mm* 2, 2-D Discrete Cosine Transform Core Processor with Variable Threshold-Voltage (VT) Scheme. IEEE Journal of Solid-State Circuits **31**(11), 1770–1779 (1996)
5. Lin, Y.S., Wu, C.C., Chang, C.S., Yang, R.P., Chen, W.M., Liaw, J.J., Diaz, C.: Leakage Scaling in Deep Submicron CMOS for SoC. IEEE Transactions on Electron Devices **49**(6), 1034–1041 (2002)
6. Liu, X., Mourad, S.: Performance of Submicron CMOS Devices and Gates with Substrate Biasing. In: The IEEE International Symposium on Circuits and Systems, vol. 4, pp. 9–12. Geneva, Switzerland (2000)
7. Mukhopadhyay, S., Mahmoodi-Meimand, H., Neau, C., Roy, K.: Leakage in Nanometer Scale CMOS Circuits. In: Proc. International Symposium on VLSI Technology, Systems, and Applications, pp. 307–312. Hsinchu, Taiwan (2003)
8. Neau, C.: Personal communication (2004)
9. Neau, C., Roy, K.: Optimal Body Bias Selection for Leakage Improvement and Process Compensation over Different Technology Generations. In: Proc. International Symposium on Low Power Electronics and Design, pp. 116 – 121. Seoul, Korea (2003)
10. Roy, K., Mukhopadhyay, S., Mahmoodi-Meimand, H.: Leakage Current Mechanisms and Leakage Reduction Techniques in Deep-Submicrometer CMOS Circuits. Proc. IEEE **91**(2), 305–327 (2003)
11. Soeleman, H., Roy, K., Paul, B.: Robust Subthreshold Logic for Ultra-low Power Operation. IEEE Transactions on Very Large Scale Integration (VLSI) Systems **9**(1), 90–99 (2001)
12. Taur, Y., Ning, T.H.: Fundamentals of Modern VLSI Devices. Cambridge University Press, New York, NY (1998)

Chapter 8
Part I: Conclusions and Future Directions

Chapter 2 described some existing leakage reduction techniques. Three main classes of techniques were discussed – power gating, body biasing and input vector control. Each of these techniques have their pros and cons and there is no one "one-size-fits-all" technique that solves the leakage problem for all designs.

In Chap. 3, we used Algebraic Decision Diagrams (ADDs) to find the histogram of leakage currents of a circuit over all input vectors. This helps us in not only finding the minimum leakage vector (MLV), but also in comparing different implementations of a circuit that have similar leakages at their MLV, but very different leakage histograms and hence different overall leakages during regular operation. The algorithm presented in Chap. 4 is also an algorithm to find the MLV of a circuit. This algorithm is however a heuristic that has much lower runtimes than the ADD-based algorithm in Chap. 3. It is hence more applicable to larger circuits. The heuristic presented used signal probabilities to guide the search for the MLV and was extended to use information on the statistical variability of leakage currents to find a MLV that reduced the mean and standard deviation of leakage. The algorithms presented in these two chapters are both useful. The advantage of the ADD based algorithm is that it yields a leakage histogram as well.

In Chap. 4, a heuristic to find the Minimum Leakage Vector (MLV) is presented. This heuristic uses signal probabilities at internal nodes to guide the search for the MLV. We also extend the heuristic to take statistical variation of leakage into account and find an optimal leakage vector that reduces the mean as well as the standard deviation of the leakage.

Chapter 5 described a new low-leakage standard cell-based ASIC design methodology – the HL methodology. The philosophy of the HL technique is to ensure that during standby operation, the supply voltage is applied across more than one *off* device and there was at least one *off* high-V_T device in the leakage path. This HL methodology requires the creation of two low-leakage variants (H and L) of each standard cell in a library. By making sure that the core of the standard cells is not touched, we ensure that the effort involved in creating these variants is not too high, thus making the approach easy to adopt. The approach assumed that the primary inputs would be set to a pre-determined value in standby. The algorithm used in our approach to convert a regular standard-cell-based design into a HL cell-based design propagated these primary input values to first determine the state of the outputs

N. Jayakumar et al., *Minimizing and Exploiting Leakage in VLSI Design*,
DOI 10.1007/978-1-4419-0950-3_8, © Springer Science+Business Media, LLC 2010

of all gates in a design during standby and then replaced them with their H or L variants. Experimental results proved that our HL methodology has better area and delay characteristics than the popular MTCMOS technique. Also, unlike MTCMOS, the leakage in our methodology is precisely estimable, after an up-front characterization of the HL library. We also investigated the feasibility of using long-channel sleep transistors instead of high-V_T sleep transistors. We find that using high-V_T transistors in the HL cells (as opposed to using long-channel sleep transistors) gives a lower leakage with a similar delay penalty. However, if mask costs are a major constraint, then using long channel length sleep transistors may be more practical. In Chap. 5 we also discussed leakage reduction in domino logic.

As we move to newer process generations, the supply voltage is expected to scale down. The threshold voltages of both high-V_T and low-V_T devices are expected to scale down as well. To keep leakage low, the threshold voltages of high-V_T devices should be kept high. While this may make the delay of the HL approach worse, the delay gets worse for only one type of transition on each gate. In the traditional MTCMOS technique, both the rising and falling transitions would get worse. Therefore, the HL technique scales better than MTCMOS with newer process technologies.

A possible modification to the HL methodology could be the sharing of the header and footer sleep transistors. This would reduce the delay considerably. This sharing of transistors could help reduce the size of the sleep transistors too. However, the area impact of this is not clear. Such a sharing of sleep transistors would require the routing of the ungated power rails as well as the routing of the power rails gated by the (now shared) sleep transistors. One possible solution would be having the H variant cells and L variant cells placed in separate (alternate) rows of the standard-cell design. The sharing of sleep transistors also opens up a little-explored avenue of research – the sizing of the sleep transistors. Even in MTCMOS, when sleep transistors are shared, the sizing of these sleep transistors is a complex problem. The authors of [6] propose an MTCMOS sleep transistor sizing algorithm, which is based on mutually exclusive discharging/charging of gates. While this technique is easily applicable to regular circuits (like a chain of inverters or decoder logic), it is hard to utilize for random logic circuits. Similarly, a precise estimation of delay is also now dependent on knowing all the mutually exclusive discharge/charge patterns. There is room for research in the area of finding the worst case (largest delay) input pattern for MTCMOS circuits and circuits that use the HL methodology with shared sleep transistors.

Another area where improvements can be made in the HL methodology is in the technology mapping phase. In our implementation, the replacing of the regular cells with their H or L variants is dependent on the primary input vector. There are several heuristics (such as those in [1–5, 7, 8]) that can be used to find a minimal leakage primary input vector for the regular standard-cell-based circuit. However, in our case once we find the best vector, we then modify the circuit (perform HL replacement). The solution we obtain is not necessarily the optimal solution, since it is quite likely that a different input vector that does not give the lowest leakage in the regular standard-cell-based circuit gives a lower leakage in the HL-cell-based circuit.

We have noticed that the HL approach worsens delay, but only for one transition for the gate. This fact can be exploited through another possible extension to the HL methodology in replacing the regular standard cells with HL cells such that the critical delay is bounded. This would involve first finding all the critical paths in a design. If a critical path utilizes the pull-up network of a gate, then we would attempt to replace that gate with a H variant. Similarly we would attempt to replace a gate with an L variant if the pull-down network of the gate is in the critical path. Yet another possible extension to the HL methodology is to create the technology mapping library so that it contains both the regular standard cells as well as their HL counterparts. We could then perform technology mapping with leakage added as one of the objectives of the mapper. The resulting circuit would contain a mix of regular standard cells and HL cells, with the HL cells used in the off-critical paths.

While most leakage reduction approaches (such as the HL and MTCMOS approaches) have a delay penalty, in Chap. 6, we presented an approach that reduces leakage while ensuring that there was no delay penalty (and in many cases a small delay improvement). We proposed an approach that combined circuit modification and input vector control at a fine-grained level. Our approach involved traversing a given circuit topologically from inputs to outputs, selectively modifying a gate so that its output (in sleep mode) is in a state that helps minimize the leakage of other gates in its transitive fanout. For this modification we developed different variants of each cell in a library, including some cells that allowed an output to be "split". While traditional input vector control only allows the primary input vector to be set so as to minimize leakage, our approach focused on circuit modifications that allowed us to not only set primary input values to a known state, but also control the logic values of internal nodes (in the standby/sleep mode). One of the key advantages of our technique is that we are able to achieve a leakage of about 30% (over input vector control alone) without a delay penalty. While other techniques such as HL or MTCMOS can achieve greater leakage savings, these techniques are orthogonal to our approach and these techniques have an associated delay penalty. Also, these approaches involve additional mask costs to create the high-V_T transistors. The approach presented in Chap. 6 does not use multiple V_T transistors and is hence less expensive to implement. Our algorithm currently replaces gates in a circuit to allow control of internal node signals (while ensuring that critical delay is not increased). If we allowed the algorithm to perform resizing of the sleep cut-off transistors used in the variants of the standard cells, we could potentially use the available slack better and achieve further leakage reductions. Sharing of the sleep cut-off transistors used is another possible improvement to the methodology. The algorithm implemented currently is a simple one that traverses a given circuit from input to output. While this makes the algorithm fast, the solution we get may not be optimal. One possible modification to our algorithm would be to first find the lowest leakage input vector, propagate this through the circuit and then target high-leakage gates and try to control their inputs.

In Chap. 7, we first present results (from a 130-nm test chip) that prove that while reverse body biasing (RBB) reduces sub-threshold leakage, the BTBT leakage component increases with greater applied RBB. Hence, there is an optimum RBB point. We presented a scheme that monitors the leakage through a representative device and finds this optimum RBB point. The scheme consists of a leakage current monitor (LCM), a programmable body bias voltage generator and digital block to interface with the LCM and the body bias voltage generator. The LCM worked by essentially converting the problem of measuring the leakage current into one of measuring the time taken for a representative leaking device to discharge (in the case of a leaking NMOS device) or charge (in the case of a leaking PMOS device) a capacitively loaded node. To cope with the large range in leakage currents, the LCM used a tunable bank of capacitors and an adjustable gate bias. The scheme presented incurred a very reasonable placed-and-routed area and also had a very small power consumption. Since the LCM presented in this chapter is small in area and not power-hungry, it could be distributed on different portions of an IC and used to monitor the leakage currents at these different points. This could be potentially useful to a designer or researcher investigating intra-die leakage variations.

The leakage reduction techniques presented in Chaps. 5– 7 are all techniques easily applicable to traditional IC design today. The techniques presented in Chaps. 5 and 6 involve some initial work in modifying or augmenting the standard-cell library. However, this task is done exactly once, upfront. There are several companies in the semiconductor industry that build standard-cell libraries. Some of them already offer low-leakage standard-cell variants as part of their libraries. The variants presented in Chaps. 5 and 6, along with the design flow and methodology to use them, could potentially be offered by these companies as part of their low-leakage standard-cell libraries. Some companies also sell blocks of logic and circuitry as Intellectual Property (IP) cores. The scheme presented in Chap. 7 is one that has potential to be offered as one such IP core.

References

1. Aloul, F., Hassoun, S., Sakallah, K., Blauuw, D.: Robust SAT-Based Search Algorithm for Leakage Power Reduction. In: Proc. Power and Timing Models and Simulation. Seville, Spain (2002)
2. Bahar, R.I., Frohm, E.A., Gaona, C.M., Hachtel, G.D., Macii, E., Pardo, A., Somenzi, F.: Algebraic Decision Diagrams and Their Applications. Formal Methods in Systems Design **10**(2/3), 171–206 (1997)
3. Gao, F., Hayes, J.: Exact and Heuristic Approaches to Input Vector Control for Leakage Power Reduction. In: Proc. International Conference on Computer-Aided Design, pp. 527–532. San Jose, CA (2004)
4. Halter, J., Najm, F.: A Gate-Level Leakage Power Reduction Method for Ultra Low Power CMOS Circuits. In: Proc. Custom Integrated Circuits Conference, pp. 475–478. Santa Clara, CA (1997)
5. Johnson, M., Somasekhar, D., Roy, K.: Models and Algorithms for Bounds on Leakage in CMOS Circuits. IEEE Transactions on Computer-Aided Design of Integrated Circuits and Systems **18**(6), 714–725 (1999)
6. Kao, J.T., Chandrakasan, A.P.: Dual-Threshold Voltage Techniques for Low-Power Digital Circuits. IEEE Journal of Solid-State Circuits **35**(7), 1009–1018 (2000)

7. Rao, R., Liu, F., Burns, J., Brown, R.: A Heuristic to Determine Low Leakage Sleep State Vectors for CMOS Combinational Circuits. In: Proc. International Conference on Computer-aided Design, pp. 689–692. San Jose, CA (2003)
8. Zhanping, C., Johnson, M., Liqiong, W., Roy, W.: Estimation of Standby Leakage Power in CMOS Circuit Considering Accurate Modeling of Transistor Stacks. In: Proc. International Symposium on Low Power Electronics and Design, pp. 239–244. Monterey, CA (1998)

Part II
Practical Methodologies for Sub-threshold Circuit Design: Exploiting Leakage Through Sub-threshold Circuit Design

While the first part of this book focused on leakage *reduction*, in the second part of this book we take a different view of leakage. Instead of minimizing leakage, we talk about *exploiting* leakage. This is achieved through sub-threshold circuit design. In the next few chapters of this book we present design methodologies that enable digital sub-threshold circuit design and operation and make it practical.

Outline of Part II

Part 2 of this book is organized as follows. In Chap. 9, we introduce the idea of operating circuits in the sub-threshold region of operation. We present exploratory studies that reveal the opportunity that sub-threshold circuits offer. We also list some of the disadvantages of sub-threshold circuit design, along with scenarios where such a methodology could be applied.

Chapter 10 presents a sub-threshold design methodology that dynamically compensates for inter and intra-die process, supply voltage and temperature (PVT) variations. This compensation is achieved by performing bulk voltage adjustments in a closed-loop fashion. Our design methodology uses a multi-level network of medium-sized Programmable Logic Arrays (PLAs) as the circuit implementation structure. The design has a global *beat clock* to which the delay of a spatially localized cluster of PLAs is "phase locked". The synchronization is performed in a closed-loop fashion, using a phase detector and a charge pump that drives the bulk nodes of the PLAs in the cluster. We demonstrate the ability of our technique to dynamically phase lock the PLA delays to the beat clock, across a wide range of PVT variations, enabling significant yield improvements. Without the approach of this chapter, the high sensitivity of the sub-threshold current to PVT variations would make sub-threshold circuit design untenable.

In Chap. 11, we first prove that while a lower voltage does result in lower *power* consumption, it does not translate to a lower *energy* consumption. In fact, we find that the optimum voltage to minimize energy consumption depends on the circuit topology. We describe a technique to find the energy optimum VDD value for a design, and show that for minimum energy consumption, the circuit may need to be operated at VDD values that are above the NMOS threshold voltage value. We study this problem in the context of designing a circuit using a network of dynamic NOR-NOR PLAs.

In Chap. 12, we propose an approach to try to reduce the speed gap between sub-threshold and traditional designs. We propose a sub-threshold circuit design approach based on asynchronous micropipelining of a levelized network of PLAs. We demonstrate that by using our approach, a design can be sped up by about 7×, with an area penalty of 47%. Further, our approach yields an energy improvement of about 4×, compared to a traditional network of PLA-based design.

Chapter 9
Exploiting Leakage: Sub-threshold Circuit Design

9.1 Overview

In the first part of this book, we discussed the problems faced due to leakage and proposed techniques to minimize leakage. In the second part of this book, we propose techniques to exploit leakage instead of minimizing it. We do this through the use of *sub-threshold circuit design*.

Because of their extreme low power consumption, sub-threshold design approaches are appealing for a widening class of applications, which demand low power consumption and can tolerate larger circuit delays.

In Sect. 9.2, the application space as well as the advantages and disadvantages of sub-threshold circuit design are presented. Section 9.2.1 details the opportunity that sub-threshold circuit design holds.

9.2 Introduction

The ever-increasing popularity of battery-powered and portable electronics underscores the importance of power consumption as a significant issue in VLSI design. There are many applications that use VLSI circuit technology where low power is essential, while the speed of operation of the device is non-critical. Let us take the example of sensor networks. It has been shown in [2, 3, 5] that sensor networks have the capability to accumulate, process and communicate information under various operating conditions. In such an application, speed is a secondary design goal, whereas low power consumption is a primary design requirement. The distributed nature of these networks, along with the need for each sensor to be maximally maintenance-free (ideally sustained by power from ambient light) further underscores the importance of low-power electronics. Further, low power consumption in these applications would reduce the amount of headroom needed for battery supplies. Also the weight of the product would be lower since smaller batteries will be sufficient to power these devices, and complex cooling solutions would not be required. Other applications that can utilize ultra-low power design techniques are wearable computers, certain portable electronic devices, implantable medical devices, etc.

N. Jayakumar et al., *Minimizing and Exploiting Leakage in VLSI Design*,

DOI 10.1007/978-1-4419-0950-3_9,

As the minimum feature size of processes continues to shrink with each successive process generation (along with the value of supply voltage and therefore V_T), leakage currents increase exponentially. On the one hand this would suggest the use of larger V_T values, but this in turn leads to slower circuits since the device (operating in linear or saturation region) has a slower turn-on when V_T is increased. Choosing a lower V_T results in lower delays but increased leakage power dissipation. Leakage power already comprises about 50% of the total power dissipation of modern designs [1, 7], so this option is not desirable either. Sub-threshold (leakage or cut-off) [9, 10] currents are hence seen as a necessary evil in traditional VLSI design methodologies. In the second part of this book, we explore techniques that turn this problem with leakage currents into an opportunity through the use of sub-threshold circuits.

Sub-threshold circuits exclusively utilize sub-threshold (leakage) currents to implement designs. This is achieved by actually setting the circuit power supply VDD to a value less than or equal to V_T. This choice results in dramatically smaller conduction currents and power at the expense of larger circuit delays. In applications such as sensor networks and wearable electronics devices, the speed of operation is not a paramount design consideration. Rather, power reduction (which translates into longer battery life, or reduced system weight resulting from the need for smaller battery packs) is a major design consideration. A practical approach to designing VLSI ICs with extremely low power consumption would be very desirable for this large and growing class of practical applications.

The advantages of a circuit design approach that utilizes sub-threshold conduction are as follows:

- Power is significantly (100–500×) lower.
- Circuits get *faster* at higher temperature [6].
- Device transconductance is an exponential function of V_{gs}, resulting in a high ratio of *on* to *off* current in a device stack. As a consequence, circuit noise margins are high.
- Delay gets worse by 10–25×, but the Power-Delay-Product (PDP) improves by 10–20×. We also show (in Chap. 11), that we can obtain an improvement in the Energy-Delay product (EDP) as well, by operating the circuit in the near-threshold region.

The disadvantages of a sub-threshold design methodology are as follows:

- I_{ds} is small, resulting in large delays.
- I_{ds} exhibits an exponential dependence on temperature, requiring circuitry to compensate for this effect.
- I_{ds} is highly dependent on process variations. For example, small changes in V_T result in large changes in I_{ds} due to the exponential dependence of I_{ds} on V_T. We therefore require circuitry to compensate for this effect as well.
- Design methodologies used today to design sub-threshold logic circuits are ad-hoc. A systematic EDA framework for the design of complex digital systems using sub-threshold logic has not been developed.

Applications such as digital wrist-watches and calculators have utilized extreme low power circuitry based on sub-threshold conduction. However, these applications are analog in nature, or implement very simple digital circuits. The design methodologies used are adhoc. A systematic EDA framework for the design of complex digital systems using sub-threshold circuits has not been developed. Our work attempts to do this and bring sub-threshold digital design into the mainstream of VLSI technology. Any practical sub-threshold methodology must address the problems of the variation of sub-threshold circuit delay with (1) temperature, (2) process variations and (3) supply voltage variations. We address these issues in the chapters in the second part of this book.

9.2.1 The Opportunity

We performed SPICE [8] experiments to compare the delay of a circuit implemented using sub-threshold CMOS logic vs. traditional CMOS logic. Our goal was to compare the delay and power values of both schemes, for a given Deep Sub-micron (DSM) process technology.

The device technologies we used were the Berkeley Predictive Technology Model [4] 0.1 μm and 0.07μm processes. For these processes, V_{T_N} and V_{T_P} are respectively 0.261 V and –0.303 V (for the 0.1 μm process) and 0.21 V and –0.22 V (for the 0.07 μm process).

Our comparison of traditional vs. sub-threshold circuit delays is shown in Table 9.1. For each process, we constructed a 21-stage ring oscillator circuit using minimum-sized inverters. From this circuit, we computed the delay, power and power-delay product for both design styles. Simulations were performed for a junction temperature of 120°C. Observe that for both the *bsim*70 and *bsim*100 processes, impressive power reductions are obtained, and the power-delay product is about 20× improved, over the traditional design style. The delay penalty can be further reduced by applying a slightly positive body bias. When the body is biased to VDD (which is set at V_T in these simulations), the delay can be brought down by a factor of 2, while the power-delay product still remains around 10× better. At this operating point, we still achieve upwards of 100× power reductions.

If V_T can be reduced further, the delay improves as indicated by the sub-threshold current equation below.

$$I_{ds}^{sub} = \frac{W}{L} I_{D0} e^{\left(\frac{V_{gs}-V_T-V_{off}}{n v_t}\right)} \left[1 - e^{-\frac{V_{ds}}{v_t}}\right]. \tag{9.1}$$

Table 9.1 Comparison of traditional and sub-threshold circuits

	Traditional Ckt			Sub-threshold Ckt ($V_b = 0\,V$)			Sub-threshold Ckt(V_b = VDD)		
Process	Delay (ps)	Pwr (W)	P-D-P (J)	Delay ↑	Power ↓	P-D-P ↓	Delay ↑	Power ↓	P-D-P ↓
bsim70	14.157	4.08e−05	5.82e−07	17.01×	308.82×	18.50×	9.93 ×	141.10×	14.43×
bsim100	17.118	6.39e−05	1.08e−06	24.60×	497.54×	20.08 ×	12.00 ×	100.96×	8.20×

Table 9.2 Sub-threshold circuit delay versus V_T for the *bsim*100 and *bsim*70 processes

*bsim*70				*bsim*100			
V_T	Delay ↑	Power↓	P-D-P ↓	V_T	Delay ↑	Power ↓	P-D-P ↓
0.180	16.15×	167.52×	10.41×	0.270	23.32×	479.85×	20.60×
0.170	14.88×	151.99×	10.09×	0.250	22.43×	464.33×	20.16×
0.160	13.78×	137.73×	9.95×	0.230	21.02×	444.23×	20.05×
0.150	13.15×	124.59×	8.86×	0.210	18.69×	400.89×	20.27×
0.140	12.43×	112.73×	9.40×	0.190	18.42×	366.28×	18.98×
0.130	12.32×	101.85×	8.02×	0.170	17.51×	323.26×	17.98×

The adjustment of V_T is easily performed during IC fabrication. We conducted experiments (for the *bsim*100 and *bsim*70 processes) to determine the reduction in delay when V_T is reduced. In these experiments, we used the same absolute value of V_T for both PMOS and NMOS devices, and operated the circuit with VDD $= V_T$. The results are reported in Table 9.2.

We note that for the *bsim*100 process, reducing V_T to 0.17 V results in a 29% delay improvement of our sub-threshold ring oscillator (at this point it is about 17.5× the delay of the traditional ring oscillator), while the power consumption remains 323× lower than that of a traditional ring oscillator (the power is about 500× lower when V_T is 0.28 V). Note that the power-delay product, an important figure of merit in circuit design, is a healthy 20× better for the sub-threshold circuit. The V_T reduction can, in practice, be achieved statically or dynamically by appropriately forward biasing the bulk node of the devices. Further, this V_T reduction can selectively be invoked for devices on the critical computation path, yielding faster designs with extremely low power consumption. Similar numbers are noted for the *bsim*70 process. The delay drops to about 12× the traditional circuit delay at $V_T = 0.13\ V$, with a 100× power improvement and a 8× improved power-delay product.

Figure 9.1 describes the trade-offs in the choice of VDD for our methodology. We show the sub-threshold current as a function of V_{gs}, for varying V_{ds} values in five steps from 0 to VDD. We show these currents with and without body bias. Note that for a given V_T, reducing VDD reduces the I_{on}/I_{off} ratio, and hence the circuit becomes less noise immune. At 0.16 V, this ratio is about 20, regardless of whether body bias is applied. Note that this means that there is no noise penalty in applying body bias. At higher voltages, this ratio improves, but less than exponentially as we move out of the sub-threshold region. Operating at a higher VDD certainly gives us larger switching currents, but the downside is that we have to switch circuit nodes over larger voltage excursions, resulting in quadratically increasing power consumption. On the other hand, operating at a lower VDD (having fixed V_T) results in lower circuit speed but much improved power reduction. For example, for the *bsim*70 process, if VDD $=$ 0.16 V (the lowest reasonable value of VDD based on noise considerations), we get a roughly 2× delay penalty and 2× power improvement from the results of Table 9.1.

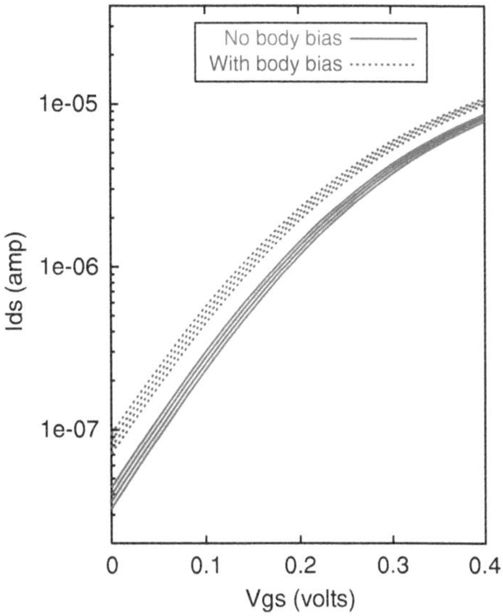

Fig. 9.1 Plot of I_{ds} versus V_{gs} (*bsim70* process)

9.3 Summary

In this chapter, we introduced the notion of exploiting leakage currents instead of minimizing them and presented experimental results that explored the opportunities that sub-threshold circuit design offers. However, sub-threshold circuits have their disadvantages and any feasible approach using sub-threshold circuits must address these disadvantages. In the next few chapters, we propose approaches that do that.

References

1. The International Technology Roadmap for Semiconductors. http://public.itrs.net/ (2003). Accessed on 12th Nov, 2003
2. The MultimodAl NeTworks of In-situ Sensors (MANTIS) Project. http://mantis.cs.colorado.edu (2004)
3. Abidi, A., Pottie, G., Kaiser, W.: Power-Conscious Design of Wireless Circuits and Systems. in Proceedings of the IEEE **88**(10), 1528–1545 (2000)
4. Cao, Y., Sato, T., Sylvester, D., Orshansky, M., Hu, C.: New Paradigm of Predictive MOSFET and Interconnect Modeling for Early Circuit Design. In: Proc. IEEE Custom Integrated Circuit Conference, pp. 201–204. Orlando, FL (2000). http://www-device.eecs.berkeley.edu/ ptm
5. Choi, S.H., Kim, B.K., Park, J., Kang, C.H., Eom, D.S.: An Implementation of Wireless Sensor Network. IEEE Transactions on Consumer Electronics **50**(1), 236–244 (2004)
6. Kanda, K., Nose, K., Kawaguchi, K., Sakurai, T.: Design Impact of Positive Temperature Dependence on Drain Current in sub-1-V CMOS VLSIs. IEEE Journal of Solid-State Circuits **36**(10), 1559–1564 (2001)
7. Mui, M., Banerjee, K., Mehrotra, A.: Power Supply Optimization in Sub-130 nm Leakage Dominant Technologies. In: Proc. 5th International Symposium on Quality Electronic Design, pp. 409–414. San Jose, CA (2004)

8. Nagel, L.: SPICE: A Computer Program to Simulate Computer Circuits. In: University of California, Berkeley UCB/ERL Memo M520 (1995)
9. Rabaey, J.: Digital Integrated Circuits: A Design Perspective. Prentice Hall Electronics and VLSI Series. Prentice Hall, Upper Saddle River, NJ (1996)
10. Weste, N., Eshraghian, K.: Principles of CMOS VLSI Design - A Systems Perspective. Addison-Wesley, Reading, MA (1988)

Chapter 10
Adaptive Body Biasing to Compensate for PVT Variations

10.1 Overview

One of the main disadvantages of their sub-threshold circuits is their extreme sensitivity to variations in power supply, temperature and processing. In this chapter, we present a sub-threshold design methodology that automatically self-adjusts for inter and intra-die process, supply voltage and temperature (PVT) variations. This adjustment is achieved by performing bulk voltage adjustments in a closed-loop fashion. The design methodology uses medium-sized Programmable Logic Arrays (PLAs) as the circuit implementation structure. Details about the structure and operation of the PLAs are presented in Sect. 10.3. The design has a global *beat clock* to which the delay of a spatially localized cluster of PLAs is "phase locked". The synchronization is performed in a closed-loop fashion, using a phase detector and a charge pump that drives the bulk nodes of the PLAs in the cluster. The details of this scheme are presented in Sect. 10.4. The experimental results presented in Sect. 10.5 demonstrate that our technique is able to dynamically phase lock the PLA delays to the beat clock, across a wide range of PVT variations, enabling the sub-threshold design methodology to be applicable in practice. We also present an analysis of the loop gain of this closed-loop adaptive body biasing technique in Sect. 10.6.

10.2 Related Previous Work

In [8–10], the authors discuss sub-threshold logic for ultra-low power circuits. They state that their approach would be useful for applications where speed is of secondary importance. In one of the two proposed approaches, they describe circuitry to stabilize the operation of their circuit across process and temperature variations. In these papers, the idea of using sub-threshold circuits was introduced from a device standpoint, and candidate compensation circuits were proposed. Also, no systematic design methodology was provided to address the multiple issues of process, temperature and supply variations within an IC die.

N. Jayakumar et al., *Minimizing and Exploiting Leakage in VLSI Design*,
DOI 10.1007/978-1-4419-0950-3_10,

In [7], the authors report a sub-threshold implementation of a multiplier. The methodology utilizes a leakage monitor and a circuit that compensates the sub-threshold current across process and temperature variations. In contrast, our approach compensates circuit *delay* directly, by phase locking it to a beat clock. In [11], a dynamic substrate biasing technique is described, as a means to make a design insensitive to process variations. The approach is described in a bulk CMOS context in contrast to our sub-threshold approach. Further, the technique of [11] matches the circuit delay to that of the critical paths (which needs to be found up-front). The dynamic biasing is not performed on a per-region basis, making it susceptible to intra-die variations.

10.3 Preliminaries: PLAs

In this section we describe the structure and operation of the PLAs used in our approach.

10.3.1 PLA Design

Consider a PLA consisting of n input variables $x_1, x_2, \ldots, x_n$, and m output variables $y_1, y_2, \ldots, y_m$. Let k be the number of rows in the PLA. A *literal* l_i is defined as an input variable or its complement.

Suppose we want to implement a function f represented as a sum of cubes $f = c_1 + c_2 + \cdots + c_k$, where each cube $c_i = l_i^1 \cdot l_i^2 \cdots l_i^{r_i}$. We consider PLAs that are of the *NOR-NOR* form. This means that we actually implement f as

$$\overline{f} = \overline{\sum_{i=1}^{k}(c_i)} = \overline{\sum_{i=1}^{k}\left(\overline{\overline{c_i}}\right)} = \overline{\sum_{i=1}^{k}\left(\overline{\overline{l_i^1} + \overline{l_i^2} + \cdots + \overline{l_i^{r_i}}}\right)}. \tag{10.1}$$

The PLA output $\overline{f}$ is a logical NOR of a series of expressions, each corresponding to the NOR of the complement of the literals present in the cubes of f. In the PLA, each such expression is implemented by *word lines*, in what is called the *AND plane*. These word lines run horizontally through the core of the PLA. Literals of the PLA are implemented by vertical-running *bit-lines*. For each input variable, there are two bit-lines, one for each of its literals. The outputs of the PLA are implemented by *output lines*, which also run vertically. This portion of the PLA is called the *OR plane*.

The PLAs in our design operate in their sub-threshold region of conduction. Figure 10.1 illustrates the schematic of the PLAs used in our design. All the PLAs in our design are of the precharged NOR NOR type and have a fixed number of inputs

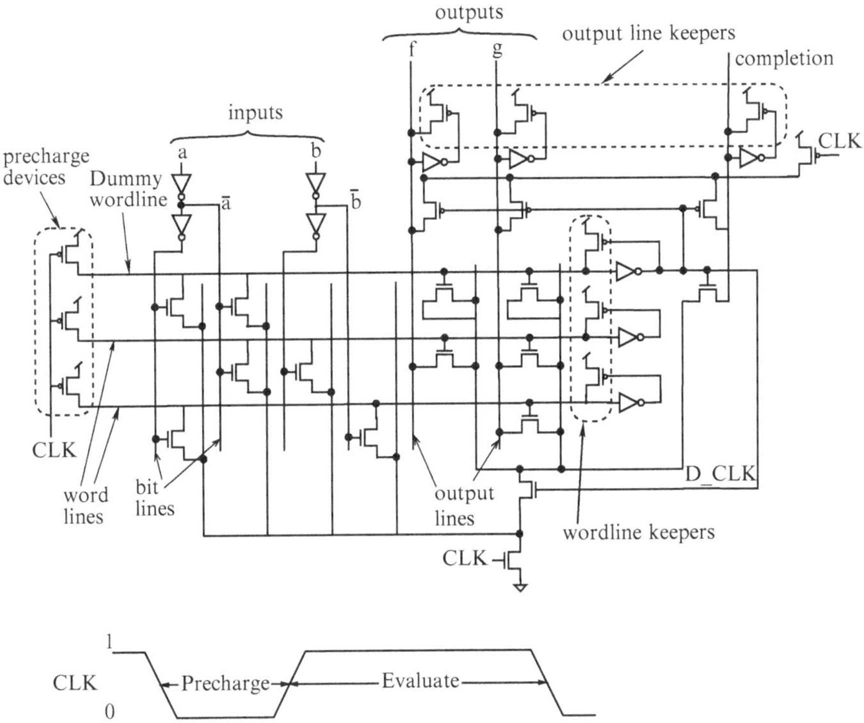

Fig. 10.1 Schematic of PLA

(12), outputs (6) and cubes (12).[1] Finally, each output of the PLAs are co-located with a negative edge triggered D flip-flop (DFF) to allow for sequential circuit support. The DFFs are not shown in Fig. 10.1. Since the PLAs evaluate in the high phase of the clock signal, the DFFs are negative edge triggered.

10.3.2 PLA Operation

The PLAs enter their precharge state when the CLK signal is low. During this time, the horizontal wordlines get precharged. A special wordline (the dummy wordline) that is the maximally loaded wordline also gets precharged. The signal on the dummy wordline is inverted to generate the delayed clock signal D_CLK. When the dummy wordline precharges (after all the other wordlines of the PLA have precharged), the delayed clock D_CLK switches low, cutting off the OR plane

[1] This was found to be a good size from a delay and area point of view for a set of benchmark circuits [3].

from GND. This delayed clock signal is also connected to PMOS pull-ups at each output line, which serve to precharge (pull-up) the output lines during the precharge phase. A special output line (which is inverted to produce the signal *completion* shown in Fig. 10.1) also gets precharged. The dummy wordline is designed to be the last wordline to switch (by making it maximally loaded among all wordlines). Similarly, the *completion* signal is also the last output signal to switch, since it is maximally loaded as well, in comparison to other outputs. The *completion* signal switching low signals the completion of the precharge operation of the PLA. In the precharged state, all the wordlines and the output lines of the PLA are precharged. Now, when the CLK signal switches high, the PLA enters the *evaluation* phase. In evaluation, if any of the vertical bitlines are high, the wordline that it is connected to gets pulled low. One of the inputs *and* its complement is connected to the dummy wordline, so that the dummy wordline switches low during *every* evaluate phase and effectively acts as a timing reference for the PLA. By design, the dummy wordline is the last wordline to switch low. When the dummy wordline switches low, it makes the signal D_CLK switch high, as a result of which the GND gating transistor in the OR plane now turns on[2]. The output lines to which wordlines that have switched low are connected, will switch low. The *completion* line that is connected to the complement of the dummy wordline is the last signal to switch high. This signals the completion of the evaluation operation. The completion signal of the PLA switches in each cycle. This signal is used to phase lock the PLA delay with the *BCLK* signal.

10.4 The Adaptive Body Biasing Solution

In this chapter, we propose a technique that uses self-adjusting body bias to *phase lock* the circuit delay to a *beat clock*. This phase locking is done for a group of spatially localized Programmable Logic Arrays (PLAs). *Therefore, inter and intra-die process variations are tackled dynamically by our approach, making our sub-threshold circuit design approach a viable means of designing extreme low power circuits.*

PLAs are chosen as the structure of choice for circuit implementation since they can be designed such that the delay is constant for all PLA outputs, regardless of the input patterns applied. This eliminates the requirement of coming up with a worst-case delay for logic, which we would require if the circuit was implemented using standard cells.

In our approach the circuit consists of a multi-level network of interconnected, medium-sized dynamic NOR-NOR PLAs[3]. Spatially localized PLAs are clustered, and each cluster of PLAs shares a common Nbulk node. This Nbulk node is

[2] Note that in the sub-threshold region a transistor is either *off* or *less off*. For the sake of simplicity, we say that an NMOS transistor is *on* when its gate is at VDD and *off* when its gate is at GND. Similarly we say a PMOS transistor is *on* when its gate is at GND and *off* when its gate is at VDD.

[3] By medium-sized PLAs, we mean PLAs that have about 5-15 inputs, 3-8 outputs, and 10-20 rows.

driven by a bulk bias adjustment circuit (one per PLA cluster), whose task it is to synchronize the delay of a representative PLA in the cluster, to a globally distributed *beat clock (BCLK)*. The beat clock is an external signal derived from the system clock. If the user would like a high speed of operation, they increase the duty cycle of *BCLK*, and all PLAs in our design speed up to synchronize to *BCLK*. Conversely, the user can reduce the frequency of *BCLK* (when the computational needs are relaxed), and the PLAs slow down and synchronize to *BCLK* again. In this way, we can implement a synchronous design methodology using sub-threshold PLAs in a manner that is insensitive to inter and intra-die processing, temperature and voltage variations.

The main problem with a sub-threshold conduction-based design approach is the strong dependency of the sub-threshold current I_{ds}^{sub} on process, temperature and voltage variations. We can see from the sub-threshold current equation that I_{ds}^{sub} has an exponential dependence on temperature. Similarly, its dependence on V_{gs} (or in other words, *VDD*) and process factors such as V_T is also exponential.

We plotted the variation of sub-threshold circuit delay[4] (for a precharged NOR-NOR PLA) against temperature, while varying various process, voltage and temperature parameters. The results are shown in Fig. 10.2. The light area represents the envelope of delays with respect to PVT variations *when no compensation was applied*. Note that the PLA delay varied by an order of magnitude. Further, in the light area of the plot, for very low temperatures (to the top and left of the Fig. 10.2) the PLA outputs did not switch at all. The parameters that were varied to compute the envelope were l_{eff} (±5% variation), V_T (±5% variation) and VDD

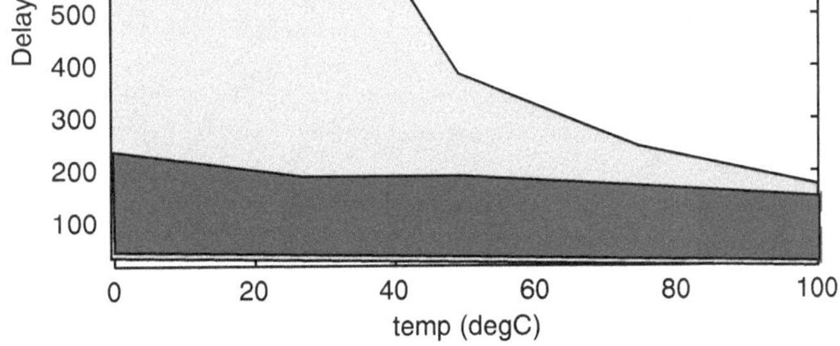

Fig. 10.2 Delay range with and without our dynamic body bias technique

[4] This is defined as the delay from the start of the evaluation phase of the computation to the time that the *completion* signal has switched.

(±10% variation). These variation values represent 3σ variation around the mean and are obtained from [12]. The dark region of Fig. 10.2 represents the PLA delay variation *after* our self-adjusting body bias technique was applied. The same variations were applied as for the light region. Note the significant reduction in the effect of PVT variations on PLA delay. Also, and importantly, *these adjustments are done in a closed-loop manner during circuit operation*. We next describe how these adjustments are made.

10.4.1 Self-Adjusting Bulk-Bias Circuit

Our self-adjusting body bias scheme controls the substrate voltage of a cluster of PLAs in a closed-loop fashion, by ensuring that the delay of a representative PLA in the cluster is phase locked to the *BCLK* signal. The phase detector and charge pump circuits for our design are shown in Fig. 10.3.

The NAND gate in this figure detects the case when the completion signal is too slow and generates low-going pulses in such a condition. These pulses are used to turn on the PMOS device of Fig. 10.3 and increase the Nbulk bias voltage, resulting in a speed-up in the PLA. The waveforms of the signals for this case are shown in Fig. 10.4. Similarly, when the completion signal is fast, the NOR gate generates pulses to turn on the NMOS device of Fig. 10.3 and hence decrease the Nbulk bias voltage. The waveforms for this situation are shown in Fig. 10.5.

Note that in general, *BCLK* is derived from *CLK*, having coincident falling edges with *CLK* but a rising edge that is delayed by a quantity D from the rising edge of *CLK*. This quantity D is the delay that we want for the evaluation of all PLAs. The value of D is computed by analyzing Fig. 10.2. We determine the largest value of delay D_{max} of the PLA for the dark region over temperatures. Now we add a suitable setup delay and phase lock error margin (in our case, we took this to be 20ns) to $D_{\max}$ to obtain D. Note that a larger margin can be chosen if we would like to be more conservative.

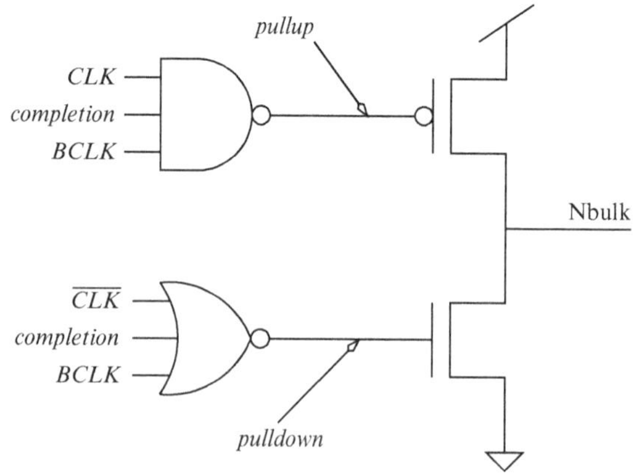

Fig. 10.3 Phase detector and charge pump circuit

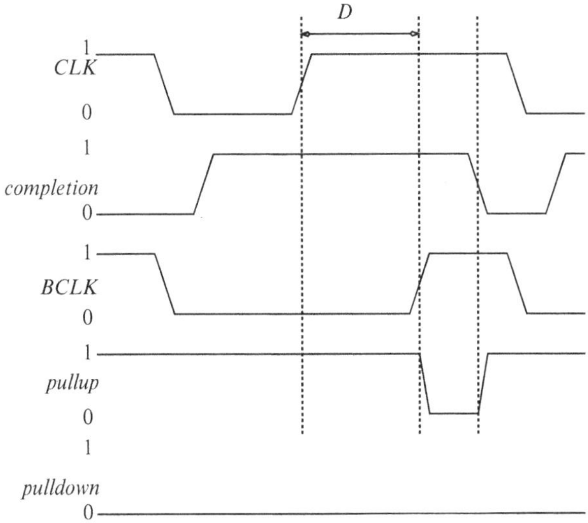

Fig. 10.4 Phase detector waveforms when PLA delay lags *BCLK*

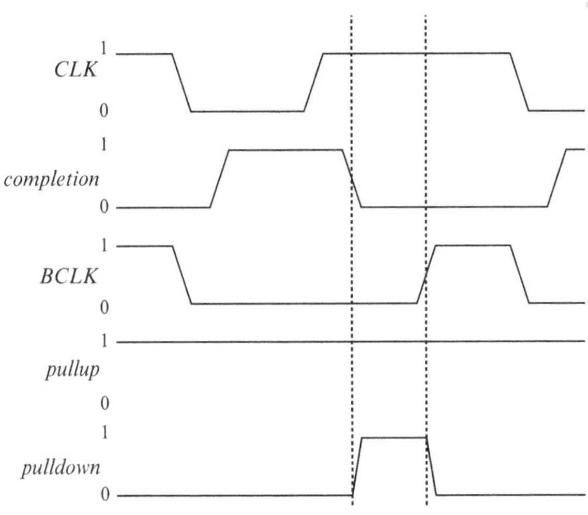

Fig. 10.5 Phase detector waveforms when PLA delay leads *BCLK*

If the completion has not occurred by the time *BCLK* rises, a downward pulse is generated on the *pull-up* signal, which forces charge into the Nbulk node, resulting in faster generation of *completion*. Note that at this time, *pull-down*, the signal that is used to bleed off charge from Nbulk is low.

The NOR gate in Fig. 10.3 generates high-going pulses to turn on the NMOS transistor when the PLA delay leads *BCLK*. These pulses drive the NMOS device in Fig. 10.3, bleeding charge out of Nbulk and thereby slowing the PLA down.

There are several observations we can make about this approach:

- Note that the PLAs in our approach operate *just fast enough* to stay synchronized with *BCLK*, thereby minimizing circuit power for a given speed of operation.
- Note that *BCLK* is used for clocking the memory elements in the design as well as for phase locking the delay of the PLA clusters.
- We do not perform bulk voltage control for PMOS devices, since there are very few PMOS devices per PLA, and they are mostly utilized for precharging purposes. It is crucial to perform bulk voltage control for NMOS devices since they are used to perform the computation during the evaluate phase of the clock.
- Sequential designs are implemented using *BCLK* as the system clock (as well as the clock used to synchronize the delays of the combinational part of the design). Additional margin is included in T_{BCLK} to account for setup delays of the memory elements and lock margin. The margin for hold times of the memory elements need not be considered since these elements are latched at the falling edge of *BCLK*.
- The distribution of the power supply and ground signals should be performed using a low-resistance supply distribution methodology such as a layout fabric [4, 5]. The power distribution network in these papers had significantly lower *iR* drops than existing power distribution approaches (up to 20× lower than traditional approaches [4]). The distribution of a sub-threshold VDD signal could be challenging, but this challenge can be averted by using a high-quality power distribution grid. Also, the switching currents in the sub-threshold design methodology are up to a couple of orders of magnitude smaller than in traditional designs, alleviating the power supply distribution problem significantly.
- We use PLAs as the circuit implementation structure because we can design them such that the delay of all outputs is constant, regardless of the input vector applied. Hence, the task of finding the critical delay path (which needs to be solved in other bulk bias control approaches such as [11]) is avoided. Also, design methodologies using a network of medium-sized PLAs were shown [5] to be a viable way to perform digital design, resulting in improved area and delay for a design. In a standard cell-based flow, there is an intervening technology mapping step, which often negates the benefits of technology-independent logic optimization. A network of PLAs on the other hand allows us to carry forward the benefits of technology-independent multi-level logic synthesis. Finally, a design implemented using such a network of PLAs can be easily mapped into a structured ASIC setting [3].

10.5 Experimental Results

We implemented our technique using PLAs as described in Sect. 10.3.1. Each cluster consisted of 1,000 spatially localized PLAs. PLAs were designed with 12 inputs, 12 rows and 6 outputs. The layout of each PLA occupied slightly over $25\,\mu \times 15\,\mu$, so each cluster was of size $0.8\,\text{mm} \times 0.5\,\text{mm}$. We simulated these PLAs using the the 65nm BSIM4 model cards from [2].

Table 10.1 Selecting the value of D

Corner	VDD	V_{Nbulk}	0°C	27°C	50°C	75°C	100°C
SS	0.18	0	n/a	685.24	376.84	251.59	169.46
		max	219.34	167.79	126.52	105.11	86.47
	0.20	0	n/a	866.15	376.12	217.01	156.98
		max	138.25	108.54	91.39	77.71	67.94
	0.22	0	n/a	n/a	360.33	204.91	148.71
		max	92.92	78.64	66.41	59.06	51.45
TT	0.18	0	254.45	168.68	139.63	105.60	82.73
		max	113.69	91.07	76.38	63.76	54.50
	0.20	0	189.59	126.91	100.19	82.22	69.11
		max	78.67	64.48	55.88	47.69	42.12
	0.22	0	135.12	102.17	82.68	63.66	59.77
		max	54.55	45.55	40.52	36.45	37.99
FF	0.18	0	88.45	67.41	61.34	46.91	40.20
		max	60.16	46.56	40.51	34.06	30.68
	0.20	0	65.41	52.19	43.11	37.60	33.48
		max	41.33	33.54	29.76	24.91	23.50
	0.22	0	47.53	40.03	34.03	30.45	25.70
		max	28.68	23.58	22.71	22.33	20.56

Table 10.1 reports the PLA delay as a function of several varying parameters. The delay is expressed as a function of l_{eff} and V_{T}, with varying VDD and V_{Nbulk}. The notation "S" indicates a slow corner, "F" indicates a fast corner, and "T" represents a typical corner. This table represents the PLA delay range that our active compensation technique can phase lock to the beat clock. Note that a "n/a" entry in Table 10.1 indicates that for the particular set of parameters, the PLA did not switch at all. The magnitude of variations for l_{eff} and V_{T} is as described earlier in this chapter, and is obtained from [12]. Note that for any process and VDD entry at any temperature, the highest speed possible is when V_{Nbulk} is maximum (i.e. set to the value of VDD for that simulation). Also, note that the ratio of the fastest to the slowest delay in this table is as high as 42:1, and our active body bias adjustment can compensate for any of these delay values.

Using Table 10.1, we can find the value of D (the amount by which we delay the rising edge of CLK to obtain $BCLK$ – please see Fig. 10.4 for illustrative purposes). We find the largest delay in the table for all rows with maximum V_{Nbulk} and add a guard-band value to this (to account for lock margin and setup margin for the memory elements). This quantity is the value of D used.

When we utilize our approach using self-adaptive body bias, the process variations described above are reduced to the dark region in Fig. 10.2. In other words, our approach is able to work for all the conditions in Table 10.1, with a delay contained in the darkened region in Fig. 10.2. The PLA delays for our approach are very tightly bounded across all these operating conditions.

Figure 10.6 describes a SPICE [6] plot of the variation of bulk voltage and PLA delay in our self-adjusting bulk bias scheme. The (higher) solid line represents the value of V_{Nbulk}, while the (lower) dotted line represents the PLA delay. Note that in

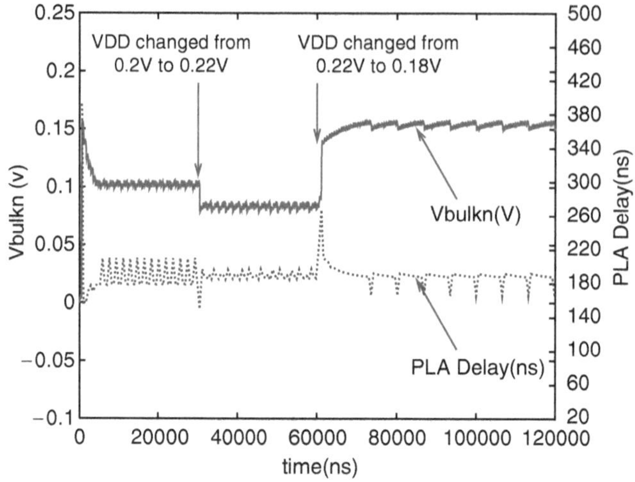

Fig. 10.6 Dynamic adjustment of PLA delay and V_{Nbulk} with VDD variation

this figure, the *VDD* value was initially 0.2 V. At time 30,000 ns, VDD was changed to 0.22 V. Note that in response to this change, our body bias adjustment circuitry modified V_{Nbulk} to a lower value in order to slow the PLAs down. At time 60,000 ns, the VDD value was changed to 0.18 V, and consequently, our bias adjustment circuit modified V_{Nbulk} to a higher value to speed up the PLAs and keep them phase locked with *BCLK*. Note that in spite of all the changes in VDD, the delay of the PLA stays tightly bounded. This simulation was done for a slow corner, at 27°C.

10.6 Loop Gain of the Adaptive Body Biasing Loop

In our scheme, we "phase lock" the delay of a representative PLA to a beat clock. We use a charge pump to adjust the body bias voltage of the PLA, which in turn controls the delay of the PLA. In principle, this scheme is a charge-pump Delay Locked Loop (DLL). An example of a traditional charge-pump DLL is shown is Fig. 10.7. In our case the representative PLA whose delay we phase lock to the beat clock takes the place of the voltage controlled delay line (VCDL) in Fig. 10.7. The phase-detector and charge pump are as shown in Fig. 10.3 The signals s_{in} and s_{out} refer to the input clock signal (or beat clock signal) and the PLA completion signal, respectively.

Based on the model shown in Fig. 10.7, we can derive the following expressions [1]:

$$s_{\mathrm{out}}(n) = s_{\mathrm{in}}(n-1) - K_{\mathrm{PLA}} V_{\mathrm{C}}(n), \tag{10.2}$$

$$V_{\mathrm{C}}(n) = \frac{s_{\mathrm{in}}(n-1) - s_{\mathrm{out}}(n-1)}{C} I_{\mathrm{p}} T, \tag{10.3}$$

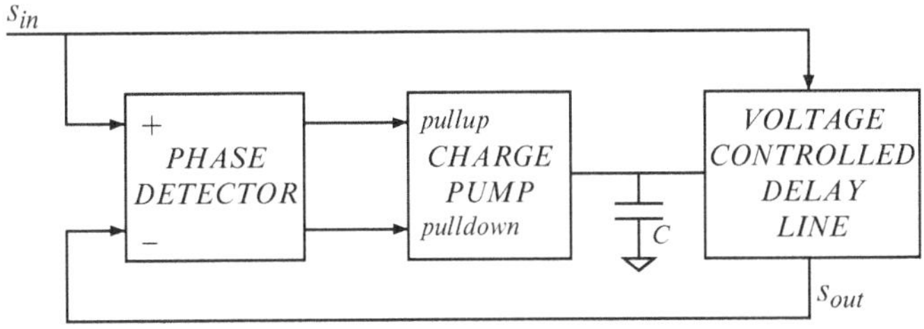

Fig. 10.7 Example of a traditional charge-pump DLL (adapted from [1])

where, $V_C(n)$ is the control voltage (body-biasing voltage) applied at the nth clock cycle, K_{PLA} is the delay gain of the PLA (ds_{out}/dV_C), I_p is the current that the charge-pump can deliver to pull-up or pull-down the control node (Nbulk node) and T is the time period of the clock. The physical meaning of (10.2) is that the arrival time of the completion signal of the representative PLA at clock cycle n is the dependent on the arrival time of the beat clock at the $n-1th$ clock cycle, the delay gain of the PLA and the control voltage at the nth clock cycle. Equation (10.3), merely states that the control voltage at the nth clock cycle is dependent on the beat clock and PLA delay at the $(n-1)th$ clock cycle, the capacitance C of the control node, the time period T and the rate at which the charge pump can pull-up and pull-down the control node.

The delay of the PLA is dependent on (inversely proportional to) the operating currents, in our case sub-threshold leakage currents. Hence, the delay of the PLA (D_{PLA}) can be written as

$$D_{PLA} = \frac{k_1}{I_{ds}}. \tag{10.4}$$

In the sub-threshold region

$$I_{ds} = \frac{W}{L} I_{D0} e^{\left(\frac{V_{gs}-V_T-V_{off}}{n v_t}\right)} \left[1 - e^{-\frac{V_{ds}}{v_t}}\right]. \tag{10.5}$$

We are only concerned with the change in I_{ds} due to change in the body-bias voltage. Hence, the expression for I_{ds} can be reduced to:

$$I_{ds} = k_2 e^{\left(\frac{V_{gs}-V_T-V_{off}}{n v_t}\right)}. \tag{10.6}$$

The body effect equation is as follows:

$$V_T = V_T^0 + \gamma\left(\sqrt{|(-2)\phi_F + V_{sb}|} - \sqrt{|2\phi_F|}\right). \tag{10.7}$$

In the above expression for V_T,
$V_{sb} = 0 - V_C$

since the source terminal is tied to GND and the bulk terminal is the control node. Substituting the above expression for V_{sb} and the expression for V_T (10.7) in the expression for I_{ds} (10.6) we get:

$$I_{ds} = k_3 e^{\left(\frac{-\gamma\left(\sqrt{|(-2)\phi_F - V_C|}\right)}{n v_t}\right)}. \tag{10.8}$$

Substituting the above expression for I_{ds} in (10.4) we get:

$$D_{PLA} = k_4 e^{\left(\frac{\gamma\left(\sqrt{|(-2)\phi_F - V_C|}\right)}{n v_t}\right)}. \tag{10.9}$$

Differentiating (10.9) with respect to V_C we get:

$$K_{PLA} = \frac{dD_{PLA}}{dV_C} = k_5 \frac{e^{\left(\frac{\gamma\left(\sqrt{|(-2)\phi_F - V_C|}\right)}{n v_t}\right)}}{\sqrt{|(-2)\phi_F - V_C|}}. \tag{10.10}$$

The expression for $s_{out}(n)$ from (10.2) can be re-written (as was shown in [1]) as:

$$s_{out}(n) = s_{in}(n-1) - K_{loop}[s_{in}(n-1) - s_{out}(n-1)]. \tag{10.11}$$

Here K_{loop} is the loop gain given by:

$$K_{loop} = \frac{K_{PLA} I_p T}{C}. \tag{10.12}$$

In the expression for loop gain K_{loop}, the current I_p is proportional to the width W of the pull-up or pull-down device. Hence, from (10.12) and(10.10) we get the expression for loop gain to be as follows:

$$K_{loop} = k_6 \frac{W T e^{\left(\frac{\gamma\left(\sqrt{|(-2)\phi_F - V_C|}\right)}{n v_t}\right)}}{C\sqrt{|(-2)\phi_F - V_C|}}. \tag{10.13}$$

The loop gain is hence proportional to the drive strength of the charge pump and inversely proportional to the capacitance of the control node. The response of our closed-loop adaptive body-biasing scheme can be adjusted using these two parameters.

10.7 Summary

Sub-threshold circuits demonstrate a dramatically reduced power consumption compared to the traditional design approaches. They are however extremely sensitive to PVT variations. In this chapter we presented a practical sub-threshold design methodology, which actively compensates for variations in supply, temperature and process. The power of our approach is its ability to adapt to inter and intra-die PVT variations, enabling a significant yield improvement.

In our design methodology, we propose using a multi-level network of medium sized Programmable Logic Arrays (PLAs) as the circuit implementation structure. Spatially localized PLAs are grouped into clusters that share a common Nbulk terminal. The design uses a global *beat clock* to which the delay of a representative PLA in this spatially localized cluster is "phase locked." Based on whether the delay of a representative PLA in any cluster leads or lags the beat clock, our approach either automatically decreases or increases the NMOS transistor bulk voltage for the cluster of PLAs. The synchronization is performed in a closed-loop fashion, using a phase detector and a charge pump that drives the bulk nodes of the PLAs in the cluster. Our results demonstrate that our technique is able to dynamically phase lock the PLA delays to the beat clock across a wide range of PVT variations. Our adaptive body-biasing scheme is in principle a charge-pump DLL. We analyzed our scheme and derived the loop gain of the system. We find that the response of the system can be tuned by adjusting the drive strength of the devices in the charge pump and the capacitance of the control (Nbulk) node.

References

1. Aguiav, R.L., Santos, D.M.: Modelling Charge-Pump Delay Locked Loops. In: Proc. International Conference on Electronics, Circuits and Systems, pp. 823–826. Pafos, Cyprus (1999)
2. Cao, Y., Sato, T., Sylvester, D., Orshansky, M., Hu, C.: New Paradigm of Predictive MOSFET and Interconnect Modeling for Early Circuit Design. In: Proc. IEEE Custom Integrated Circuit Conference, pp. 201–204. Orlando, FL (2000). http://www-device.eecs.berkeley.edu/ ptm
3. Jayakumar, N., Khatri, S.: A METAL and VIA Maskset Programmable VLSI Design Methodology Using PLAs. In: Proc. IEEE/ACM International Conference on Computer Aided Design, pp. 590–594. San Jose, CA (2004)
4. Khatri, S., Mehrotra, A., Brayton, R., Sangiovanni-Vincentelli, A., Otten, R.: A Novel VLSI Layout Fabric for Deep Sub-Micron Applications. In: Proc. Design Automation Conference. New Orleans, LA (1999)
5. Khatri, S.P., Brayton, R.K., Sangiovanni-Vincentelli, A.: Cross-talk Immune VLSI Design Using a Network of PLAs Embedded in a Regular Layout Fabric. In: Proc. IEEE/ACM International Conference on Computer Aided Design, pp. 412–418. San Jose, CA (2000)
6. Nagel, L.: SPICE: A Computer Program to Simulate Computer Circuits. In: University of California, Berkeley UCB/ERL Memo M520 (1995)
7. Paul, B., Soeleman, H., Roy, K.: An 8X8 Sub-Threshold Digital CMOS Carry Save Array Multiplier. In: Proc. European Solid State Circuits Conference, pp. 377–380. Villach, Austria (2001)
8. Soeleman, H., Roy, K.: Ultra-low Power Digital Subthreshold Logic Circuits. In: Proc. International Symposium on Low Power Electronic Design, pp. 94–96. San Diego, CA (1999)
9. Soeleman, H., Roy, K.: Digital CMOS Logic Operation in the Sub-threshold Region. In: Proc. Tenth Great Lakes Symposium on VLSI, pp. 107–112. Chicago, IL (2000)
10. Soeleman, H., Roy, K., Paul, B.: Robust Subthreshold Logic for Ultra-low Power Operation. IEEE Transactions on Very Large Scale Integration (VLSI) Systems **9**(1), 90–99 (2001)

11. Tschanz, J., Kao, J., Narendra, S., Nair, R., Antoniadis, D., Chandrakasan, A., De, V.: Adaptive Body Bias for Reducing Impacts of Die-to-Die and Within-die Parameter Variations on Microprocessor Frequency and Leakage **37**, 1396–1402 (2002)
12. Zarkesh-Ha, P., Mule, T., Meindl, J.D.: Characterization and Modelling of Clock Skew with Process Variation. In: Proc. IEEE Custom Integrated Circuits Conference, pp. 441–444. San Diego, CA (1999)

Chapter 11
Optimum VDD for Minimum Energy

11.1 Overview

Operating circuits in the sub-threshold region or near the sub-threshold design can yield extremely low power circuits. However, for most applications that require ultra-low power, the lowest power solution is not necessarily the optimal solution from a minimum energy point of view. In this chapter, we describe a technique to find the energy optimum VDD value for a design, and show that for minimum energy consumption, the circuit may need to be operated at VDD values that are slightly higher than the NMOS threshold voltage value. We study this problem in the context of designing a circuit using a network of dynamic NOR-NOR PLAs.

In Sect. 11.3, we present related previous work. Some preliminaries and assumptions in this chapter are mentioned in Sect. 11.4 while the experiments that demonstrate how the optimum VDD was calculated are discussed in Sect. 11.5.

11.2 Introduction

Power is minimized by operating the design at a lower voltage. However, a practical approach to designing VLSI ICs with minimum *energy* consumption would be very desirable for a large and growing class of practical applications. While it has been shown that power consumption is lower for lower voltages, the energy consumption per operation (i.e. the energy consumption for a logic gate to perform one computation) is not necessarily lower for lower VDDs. This is due to the fact that since switching times are longer, the power consumption over that longer switching period causes a greater energy consumption. In this chapter we describe an approach to finding the optimal VDD value for energy minimization. We assume that the circuits in question can be operated over a range of VDD values (including sub-threshold and super-threshold values of VDD).

We address the problem of finding the optimal VDD value for minimum energy consumption in a design scenario where a design is implemented using a network of medium-sized Programmable Logic Arrays (PLAs) [5]. This design approach was

N. Jayakumar et al., *Minimizing and Exploiting Leakage in VLSI Design*,
DOI 10.1007/978-1-4419-0950-3_11,

shown recently to be suitable for implementing structured ASICs with a low-NRE cost [4]. Also, it was indicated in a recent keynote talk [11] that PLAs are strong contenders as the circuit implementation structures of choice in future designs.

11.3 Related Previous Work

There has been some recent research in the area of sub-threshold operation [10, 12–15] for standard-cell based designs. These designs consume extremely low power. However, as has been pointed out in [1, 3, 16], while the optimum VDD for minimum power is the lowest possible VDD value, the optimum VDD for minimum energy can be higher, especially in situations where the static power consumption is comparable to the dynamic power consumption.

In [3], a first-order model of the energy-delay product (EDP) is reported . Using this model, the authors find the optimum VDD and body bias point for CMOS circuits operating in strong inversion. In [1], the authors examine the effects of device sizing on energy for standard-cell based circuits operating in the sub-threshold region. In [16], the performance and energy dissipation contours for CMOS circuits operating in the sub-threshold region are presented, to help find the optimum VDD and threshold voltage. The authors of [16] also point out that these contours change depending on the switching probabilities of the circuit nodes. Hence, the optimum VDD is heavily dependent on the type of circuit. Similarly, in [17], the authors describe theoretical and practical considerations for energy minimization in dynamic voltage scaled systems, allowing for sub-threshold operation.

In this work, as in [16, 17], we attempt to find the optimum VDD that minimizes energy for a circuit. However, in contrast to prior approaches, we use fixed-size dynamic NOR-NOR PLAs instead of standard cells as the circuit implementation approach. One of the advantages of this design choice is that it allows us to come up with the optimum VDD for any design with just the knowledge of the logic depth (in terms of the number of PLAs) of the design and the energy characterization data of a *single* PLA. This is not feasible for previous standard cell-based approaches. As a consequence, our approach is applicable to a network of PLA-based designs, including structured ASICs [4] implemented under this methodology. Further, in contrast to the approaches of [16, 17], our network of PLA-based approach has an energy consumption that is *highly predictable and largely independent of the input vector applied to the design*. This fact arises from the regularity inherent in the PLAs. Also, in contrast with [17], we study the dependence of the optimal VDD point on temperature.

The ability to find the optimum VDD for a network of PLA circuits using the characterization data from just a single PLA allows us a significant advantage in a practical design setting. We can find the optimum VDD for a circuit by only knowing its topological depth in terms of number of PLAs. We do not need to know any additional design details.

11.4 Preliminaries

The aim of this work is to explore how energy can be minimized in a circuit designed using a network of precharged NOR-NOR PLAs. Towards this end, we first explore the effect (in terms of power, delay and energy consumption) of changing VDD and Vbulkn (the body bias of NMOS devices in the PLA) for a single PLA and then use this information to help find an optimum VDD value for a circuit designed using these PLAs.

The PLA we use is a precharged NOR-NOR PLA (similar to the ones used in [5–8] and Chap. 10). The structure and operation of the PLA is presented here again for the reader's convenience. The PLAs we consider have a fixed number of inputs (12), outputs (6) and rows (12).[1]

11.4.1 Operation of the PLA

The structure of the PLA used for the experimental results in this chapter is shown in Fig. 11.1. When the CLK signal is low (logic-0), the PLA enters the *precharge* phase. During this time, the horizontal wordlines get precharged. A special wordline (the dummy wordline), which is the maximally loaded wordline, also gets precharged. This forces the signal D_CLK to go low, cutting off the OR plane from GND and causing the output lines to also get precharged. A special output line (marked *completion* in Fig. 11.1) also gets precharged. The dummy wordline is designed to be the last wordline to switch (by making it maximally loaded among all wordlines). Similarly, the completion line is also the last output line to switch, since it is maximally loaded as well, in comparison to other outputs. The *completion* line switching high signals the completion of the precharge operation of the PLA. In the precharged state, all the wordlines and the output lines of the PLA are precharged. Now, when CLK switches high, the PLA enters the *evaluation* phase. In evaluation, if any of the vertical bitlines are high, the wordline that it is connected to gets pulled low. One of the inputs *and* its complement are connected to the dummy wordline, so that the dummy wordline switches low during *every* evaluate phase. By design, the dummy wordline is the last wordline to switch low. This makes the signal D_CLK go high, as a result of which the GND gating transistor in the OR plane now turns on. The output lines to which wordlines that have switched low are connected will switch low. The *completion* line, which is connected to the complement of the dummy wordline, is the last line to switch low. This signals the completion of the evaluation operation.

A circuit implemented using a network of PLAs operates as follows. All PLAs precharge when the global clock signal is low. When the global clock is high, the

[1] We fix these values for each PLA in the design so as to be able to utilize the PLAs in a structured ASIC setting, allowing for a low-NRE design approach.

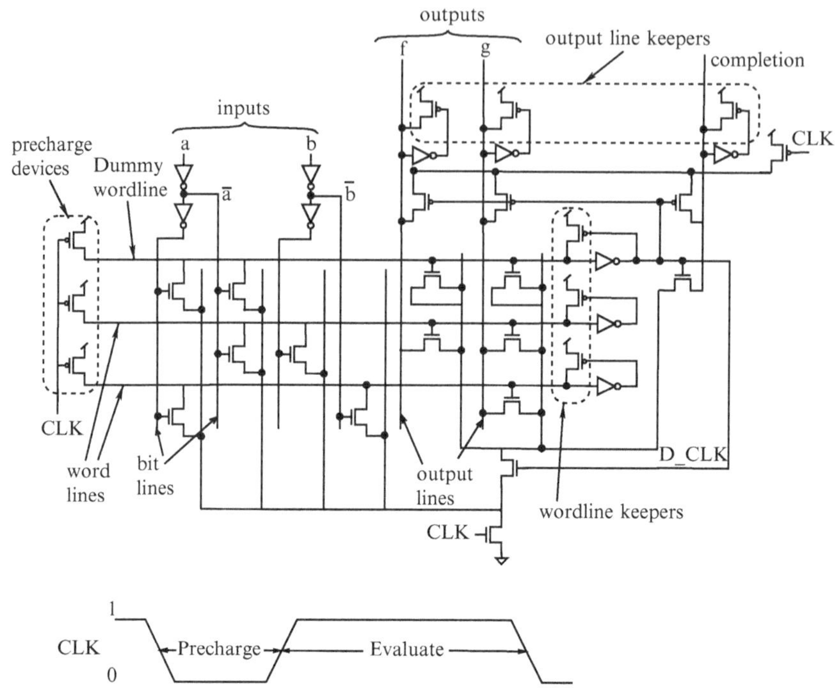

Fig. 11.1 Schematic of PLA

PLAs evaluate. The evaluation condition of a PLA of topological depth i is the global clock, gated by the completion signal of the slowest PLA among the PLAs of level $i - 1$.

11.4.2 Some Definitions

Since the PLAs used are of fixed size, the characterization of a single PLA provides enough information to estimate the delay, power and energy consumption of a circuit built using these PLAs as building blocks. The regularity of the PLAs, which allows us to infer circuit level delay, power and energy estimates from those of a single PLA, is an additional advantage of this design approach.

We divide the modes of operation of the PLA into four different phases in order to characterize it more easily. These are the *Precharging mode*, the *Precharged mode*, the *Evaluating mode* and the *Evaluated mode*. This partitioning of modes is shown in Fig. 10.1. The *Precharging mode* refers to the period of operation during which the PLA is precharging. In this mode, all wordlines and output lines get pulled high. The *Precharging time*, T_{pchg} is defined to be the time from which the clock starts to go low (1% below VDD) to the time when the completion signal of the PLA reaches logic high (within 1% of VDD). Similarly the *Evaluating mode* refers to the period when the PLA is evaluating. This is the period during which the wordlines and the output lines are switching low (depending on the inputs to the PLA). The *Evaluating time*, T_{eval} is defined to be the time from when the clock starts to go high (1% of VDD above GND) to the time when the completion line reaches logic low (reaches within 1% of VDD above GND). The *Precharged mode* refers to the period when the PLA is precharged and is idle (waiting for the clock to go high to start evaluation). Similarly, the *Evaluated mode* refers to the mode of operation where the PLA has completed evaluation and is idle (waiting for the clock line to go low to start the next precharge operation). The power consumed in the Precharging and the Evaluating modes is classified as *dynamic power consumption*, while the power consumption in the Precharged mode and the Evaluated mode is classified as *static power consumption*. Note that the static power consumption includes power consumption due to all forms of leakage currents [sub-threshold leakage, gate leakage and gate induced drain leakage (GIDL)]. Let $EvalEnergy_{dyn}$ denote the energy consumption in the Evaluating mode, $PchgEnergy_{dyn}$ denote the energy consumption in the Precharging mode, $EvalPwr_{sta}$ denote the power dissipated in the Evaluated mode and $PchgPwr_{sta}$ denote the power dissipated in the Precharged mode. The *evaluation delay* is defined as the difference between the time instant the clock line voltage crosses VDD/2 (clock line rising) and the instant when the completion line crosses VDD/2 (completion line falling). In the operation of the PLA, the evaluation delay is the critical delay of the PLA.

11.5 Experiments

For our simulations, we used Spice3 [9] with 65-nm BSIM4 [2] model cards. The threshold voltages for our devices were $V_{T_n} = 0.22\,\text{V}$ and $V_{T_p} = -0.22\,\text{V}$. In this section we will discuss the results of these simulations and describe a methodology to find an optimum VDD value for a circuit, so as to minimize energy consumption. The range of VDD values that are of interest vary from slightly below V_{T} to a few 100mV above V_{T}. Hence, we refer to our operating voltage range as *near-threshold*.

Figure 11.2 shows the plot of power for the PLA (for each of the four modes) for an operating temperature of 25°C. The power is plotted at varying VDD levels. The plot also shows the dependence of the evaluation delay on VDD. Not surprisingly, the delay increases at lower voltages while power dissipation is reduced. Similar results were seen at other temperatures and different Vbulkn values.

Figure 11.3 shows plots of the power dissipated for the different modes with varying Vbulkn at different VDD values. The temperature was fixed at 25°C. The plots for other temperatures are similar. The evaluation delay variation with Vbulkn is also shown. As can be seen from these plots, at low voltages (especially at sub-threshold voltages), a forward body bias of 0.2 V can give more than a 2× speedup but with a proportionate power penalty. Forward body biasing helps reduce delay for higher voltages as well, but the effect is greater at low/sub-threshold voltages.

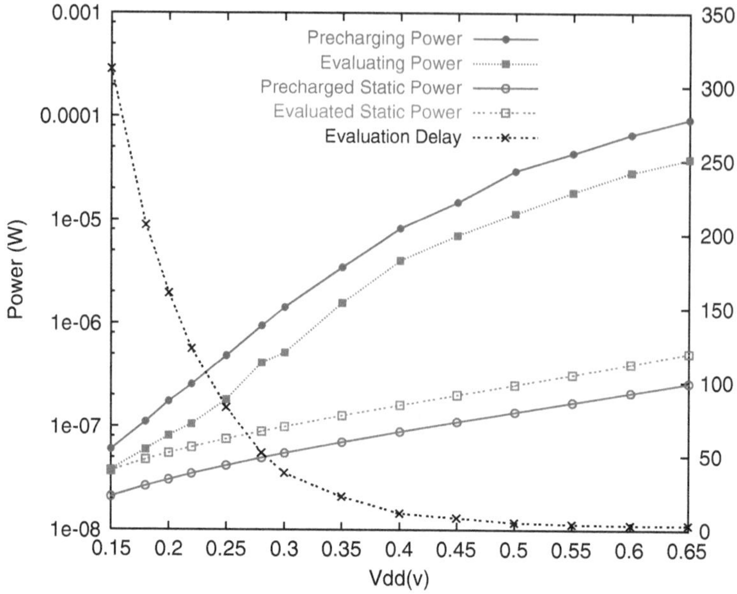

Fig. 11.2 Power dissipated, delay in the four modes with varying VDD (Vbulkn = 0 V)

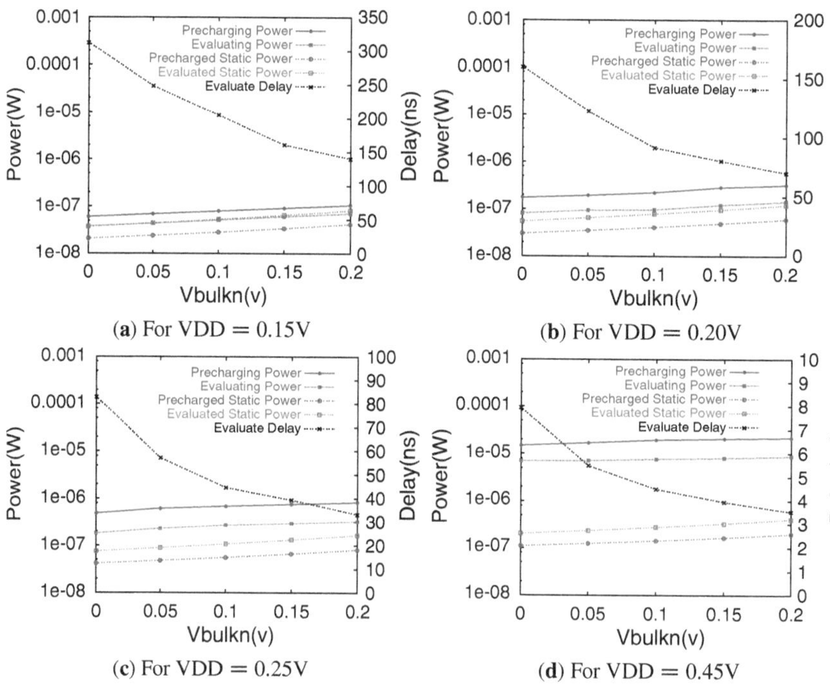

Fig. 11.3 Power and delay in all four modes with varying Vbulkn

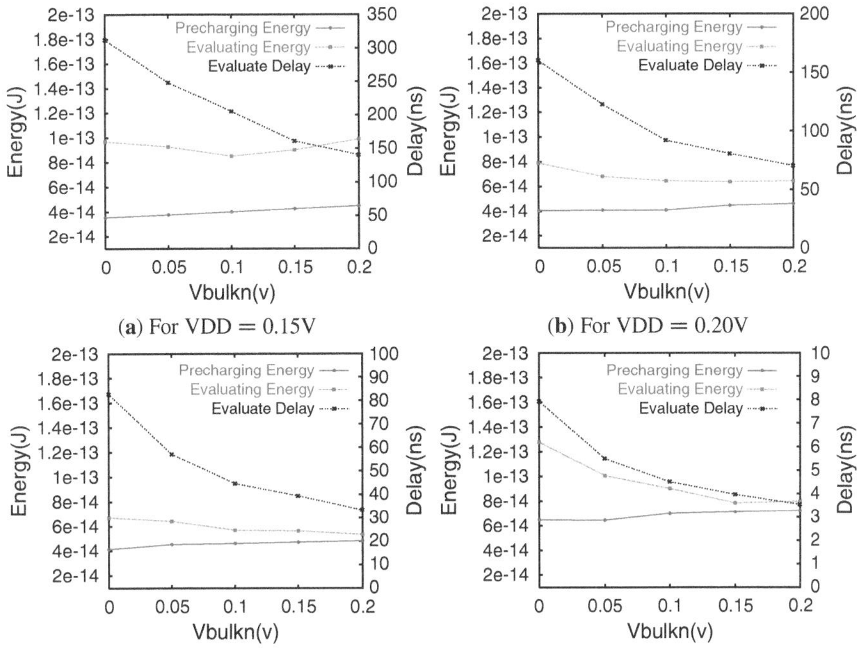

(**a**) For VDD = 0.15V (**b**) For VDD = 0.20V (**c**) For VDD = 0.25V (**d**) For VDD = 0.45V

Fig. 11.4 Energy consumption and delay in the two dynamic modes, with varying Vbulkn

Figure 11.4 shows plots of the energy consumption with varying Vbulkn for different VDD values at a temperature of 25°C. These plots indicate that even with the increase in power due to forward body biasing, the energy consumption does not increase significantly and can in fact decrease with increasing forward body bias. This would suggest that a forward body bias helps since it decreases delay without an energy penalty. However, rather than drive this body bias voltage with a fixed value, it is suggested that this body-bias control be used adaptively as suggested in Chap. 10 to control the speed of the PLA circuit over varying process corners and temperatures. This is because devices in the sub-threshold region of operation are more susceptible to temperature and process variations.

Figure 11.5 plots the energy consumption in the evaluating period and in the precharging period of the PLA. The evaluation delay is also shown. Note that the evaluation delay is measured at the VDD/2 crossing points. This delay is smaller than the evaluating time T_{eval} (see definitions in Sect. 11.4.2).

Intuitively, for minimum energy consumption, no time should be spent in the idle modes (Precharged mode and Evaluated mode). However, in a circuit constructed using a network of PLAs of fixed size, some of the PLAs may have to remain in the Precharged state or in the Evaluated state for a certain period of time. This duration is dependent on the topological depth of the network of PLA circuit (as we shall see in Sect. 11.5.1).

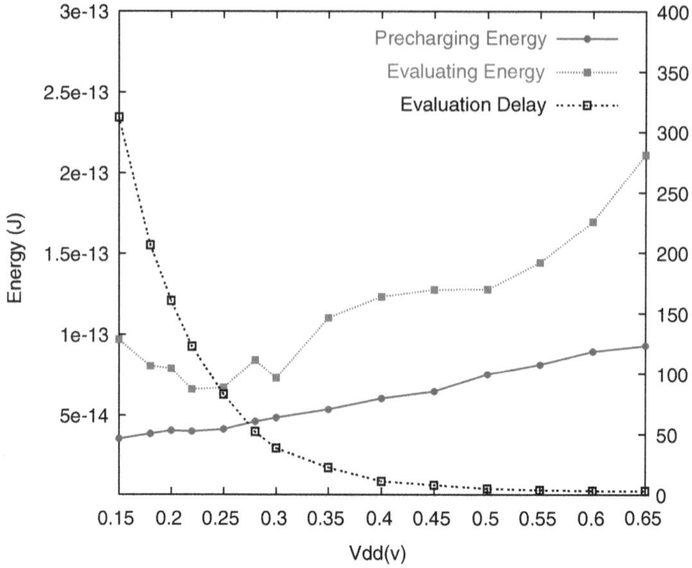

Fig. 11.5 Energy consumption, delay in the two dynamic modes with varying VDD (Vbulkn = 0 V)

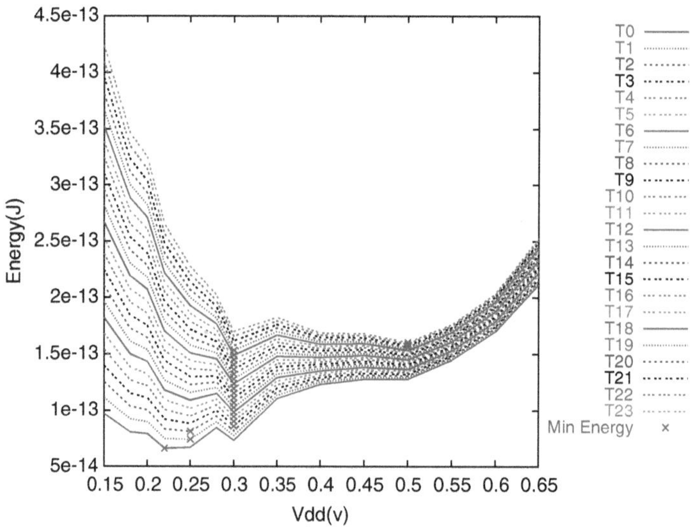

Fig. 11.6 Energy consumption over different activity factors (Vbulkn = 0 V)

The evaluation energy consumption is plotted against VDD in Fig. 11.6. The different curves denote the different ratios of evaluating time to time spent in the evaluated state. T0 represents only the evaluating energy consumption (no time spent and hence no energy consumed in the evaluated state). T1 denotes the sum of energy consumption during the evaluating period (dynamic energy consumption in the evaluating period) and energy consumption in the evaluated state for a period equal to the evaluating time. In other words, the curve T1 plots $energy = EvalEnergy_{dyn} + (T_{\text{eval}} \times EvalPwr_{sta})$. Similarly the curve T2 plots $energy = EvalEnergy_{dyn} + (2 \times T_{\text{eval}} \times EvalPwr_{sta})$, and so on. In essence, Fig. 11.6 plots the energy consumption for different activity factors i.e. the ratios of time spent in the evaluating state to the time spent in the static (idle) evaluated state.

As can be seen from the plot in Fig. 11.6, as more time is spent in the static (idle) modes (i.e. in regions where static power is dissipated), the optimum VDD value (which minimizes energy) tends to shift to higher values.

11.5.1 Energy Estimation for a Circuit of PLAs

The operation of a combinational circuit designed with a network of multi-level fixed size PLAs is as follows. Assume that the circuit has a topological depth D. In other words, the longest path between any circuit input and any circuit output traverses D PLAs. All PLAs are precharged simultaneously. Once all the PLAs are precharged, the global clock line goes high for all the PLAs. The PLAs evaluate in a domino fashion, starting with PLAs of topological level 1 and proceeding to PLAs of topological level D. The local clock of the level 1 PLAs is ungated, so level 1 PLAs evaluate as soon as the global clock goes high. The local clock of level i PLAs is gated by the completion signal of a representative level $i-1$ PLA. As a result, once the *completion* signal of level $i-1$ PLAs goes low, level i PLAs begin evaluation. In this manner, the evaluation of PLAs proceeds in topological levelization order.

An example of such a series of four PLAs is shown in Fig. 11.7. PLA1 receives its input externally. PLA2 may receive its inputs externally and/or from PLA1. PLA3 may receive its inputs externally, from PLA2 and/or from PLA1. PLA4 may receive its inputs externally, from PLA3, and/or from PLA2 and PLA1. Note that since the PLAs are of fixed size, each of the PLAs have the same evaluating time.

All four PLAs are precharged at the same time. This operation is completed in time T_{pchg}. Next PLA1 evaluates, taking time T_{eval} to do so. Once the outputs of PLA1 are ready, the next PLA, PLA2 evaluates. Once PLA2 completes evaluation, PLA3 starts evaluating and after PLA3 completes evaluation, PLA4 evaluates. After PLA4 has completed its evaluation, the circuit is again precharged to get ready for the next set of inputs. As can be seen from the timing diagram in Fig. 11.7, PLA1 is in the evaluated state for a period $t_6 - t_3 = 3 \cdot T_{\text{eval}}$. During this period, the energy consumption by PLA1 $= EvalPwr_{sta} \times 3 \cdot T_{\text{eval}}$, since the energy consumption during this period is due to the static power consumption in the Evaluated state. Similarly, we find that PLA2 is in the evaluated state for a period $= 2 \cdot T_{\text{eval}}$, while PLA3 is in the evaluated state for a period $= T_{\text{eval}}$. Figure 11.7 also reveals that PLA4 is in the Precharged state for the period $t_5 - t_2 = 3 \cdot T_{\text{eval}}$ and during this period the energy

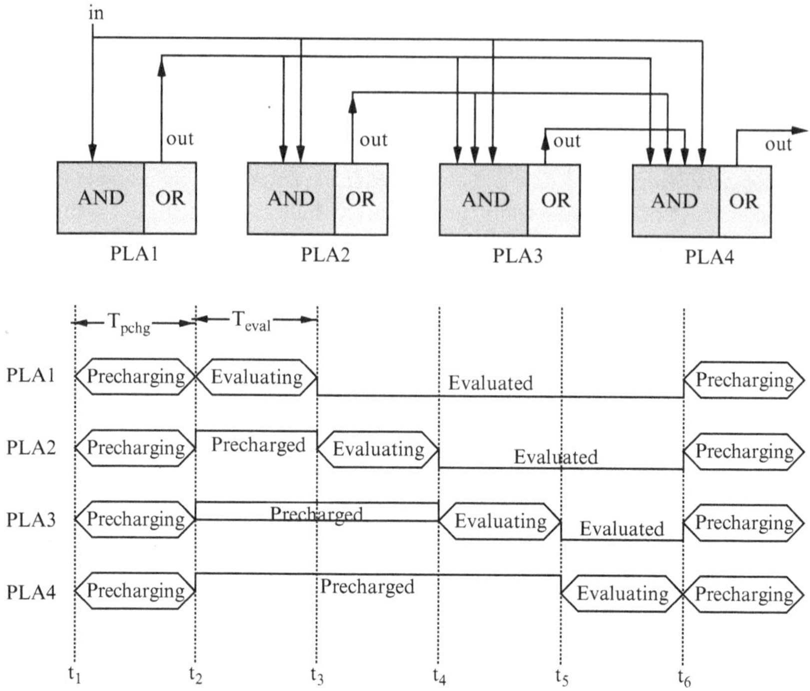

Fig. 11.7 Circuit built as a series of four PLAs

consumption is given by $PchgPwr_{sta} \times 3 \cdot T_{\text{eval}}$ since it is the static power consumption in the Precharged state that contributes to the energy consumption during this time. Similarly, we find that PLA3 and PLA2 are in the precharged state for the durations of $2 \cdot T_{\text{eval}}$ and T_{eval}, respectively.

Hence, for a PLA in a circuit of topological depth D (in terms of number of PLAs), we can estimate the energy consumption for a PLA at depth k as follows:

$$\begin{aligned} Energy &= PchgEnergy_{dyn} + EvalEnergy_{dyn} \\ &\quad + [PchgPwr_{sta} \times T_{\text{eval}} \times (k-1)] \\ &\quad + [EvalPwr_{sta} \times T_{eval} \times (D-k)] \end{aligned} \tag{11.1}$$

If the circuit consists of n PLAs connected in a chain as in Fig. 11.7, the total energy consumption for all n PLAs is given by:

$$\begin{aligned} Energy = {} & [(PchgEnergy_{dyn} + EvalEnergy_{dyn}] \times D) + [(D \times (D-1)/2) \\ & \times (EvalPwr_{sta} + PchgPwr_{sta}) \times T_{eval}]. \end{aligned}$$

If the network of PLAs is not structured like a chain, the total energy is computed by summing the energies for each PLA, from 1.

Using this equation, we plotted the energy consumption for network of PLA circuits with different topological depths, with varying VDD. This plot is shown for different temperatures for circuits up to a logic depth of 24 (labeled Depth0 through Depth23) in Figs. 11.8–11.11.

We find that while power is lower at lower voltages, there is greater energy consumption per cycle of operation at very low voltages, since the PLA takes longer to switch. This gets worse when the PLA is idle for longer periods (which is inevitable in PLAs circuits with large topological depths). In fact, we find that for such circuits, a higher VDD gives better energy consumption per cycle. Also, we have experimentally validated that the optimum VDD selection is independent of the logic function being implemented, provided the topological depth remains unchanged. Another observation that can be made is that as leakage becomes a larger component of the total power dissipation, the optimum VDD value also increases (in order to reduce the idle time of each PLA). Hence under a forward body bias voltage (which would decrease V_T and thereby increase leakage), the optimum VDD increases.

The optimal value of VDD for minimum energy is between V_T and about $1.5V_T$ for low temperature operation, while it increases to between $1.5V_T$ and $2.5V_T$ for higher temperatures. This suggests that for extreme low power applications such as sensor networks, where the ambient temperature conditions may vary significantly, special temperature compensation circuitry would be required.

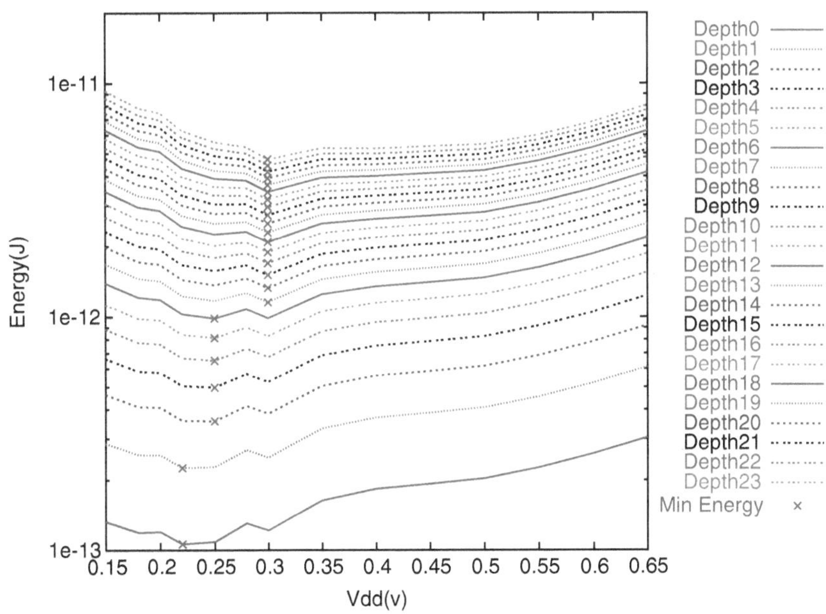

Fig. 11.8 Total energy consumption per cycle for different logic depths at 25°C (Vbulkn = 0 V)

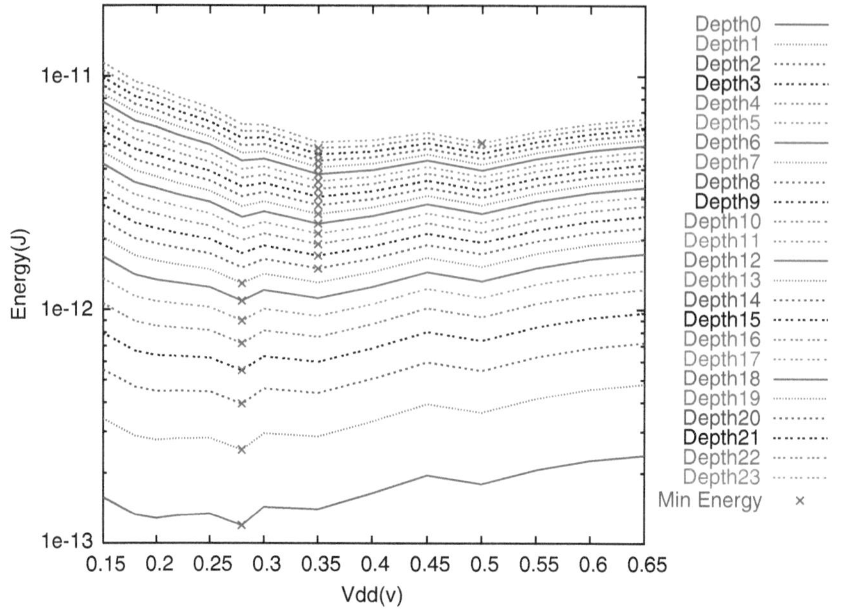

Fig. 11.9 Total Energy consumption per cycle for different logic depths at 50°C (Vbulkn = 0 V)

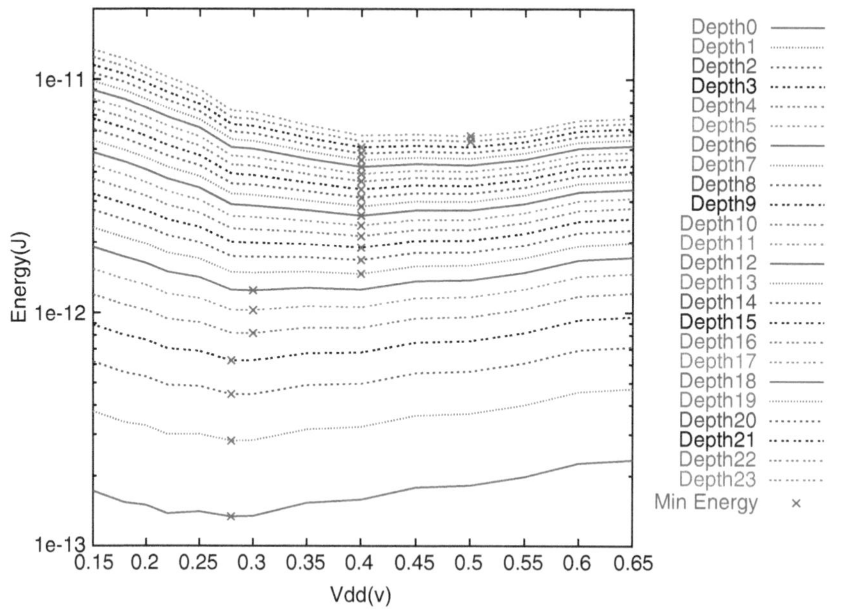

Fig. 11.10 Total Energy consumption per cycle for different logic depths at 75°C (Vbulkn = 0 V)

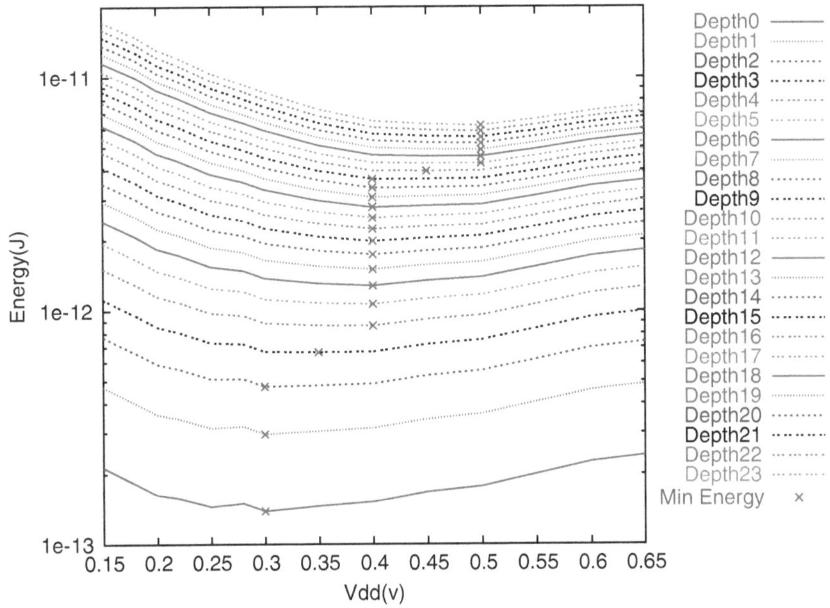

Fig. 11.11 Total energy consumption per cycle for different logic depths at 100°C (Vbulkn = 0 V)

11.6 Summary

In recent times, there has been a significant growth in applications for battery-powered portable electronics, as well as low power sensor networks. For such systems, energy minimization is a dominant design constraint, whereas circuit speed is a secondary requirement. In this chapter, we focused on finding the optimal VDD value for energy minimization of circuits that are implemented in a network of PLA design approach. We find that the optimal VDD value for such designs is close to V_T for circuits with low topological depth, but increases to about $2.5V_T$ for circuits with large topological depth and increasing temperature.

References

1. Calhoun, B.H., Wang, A., Chandrakasan, A., Kosonocky, S.: Device Sizing for Minimum Energy Operation in Subthreshold Circuits. In: Proc. IEEE Custom Integrated Circuits Conference, pp. 95–98. Orlando, FL (2004)
2. Cao, Y., Sato, T., Sylvester, D., Orshansky, M., Hu, C.: New Paradigm of Predictive MOSFET and Interconnect Modeling for Early Circuit Design. In: Proc. IEEE Custom Integrated Circuit Conference, pp. 201–204. Orlando, FL (2000). http://www-device.eecs.berkeley.edu/ ptm
3. Gonzalez, R., Gordon, B.M., Horowitz, M.A.: Supply and Threshold Voltage Scaling for Low Power CMOS. IEEE Journal of Solid-State Circuits **32**(8), 1210–1216 (1997)

4. Jayakumar, N., Khatri, S.: A METAL and VIA Maskset Programmable VLSI Design Methodology Using PLAs. In: Proc. IEEE/ACM International Conference on Computer Aided Design, pp. 590–594. San Jose, CA (2004)
5. Khatri, S., Mehrotra, A., Brayton, R., Sangiovanni-Vincentelli, A., Otten, R.: A Novel VLSI Layout Fabric for Deep Sub-Micron Applications. In: Proc. Design Automation Conference. New Orleans, LA (1999)
6. Mo, F., Brayton, R.: River PLAs: A Regular Circuit Structure. In: Proc. Design Automation Conference, pp. 201–206. New Orleans, LA (2002)
7. Mo, F., Brayton, R.: Whirlpool PLAs: A Regular Logic Structure and Their Synthesis. In: Proc. IEEE/ACM International Conference on Computer Aided Design, pp. 543–550. San Jose, CA (2002)
8. Mo, F., Brayton, R.: PLA-Based Regular Structures and Their Synthesis. IEEE Transactions on Computer-Aided Design of Integrated Circuits and Systems **22**(6), 723–729 (2003)
9. Nagel, L.: SPICE: A Computer Program to Simulate Computer Circuits. In: University of California, Berkeley UCB/ERL Memo M520 (1995)
10. Paul, B., Soeleman, H., Roy, K.: An 8X8 Sub-Threshold Digital CMOS Carry Save Array Multiplier. In: Proc. European Solid State Circuits Conference, pp. 377–380. Villach, Austria (2001)
11. Rabaey, J.: Design at the End of the Silicon Roadmap. Keynote Talk, Asia and South Pacific Design Automation Conference (2005)
12. Soeleman, H., Roy, K.: Ultra-low Power Digital Subthreshold Logic Circuits. In: Proc. International Symposium on Low Power Electronic Design, pp. 94–96. San Diego, CA (1999)
13. Soeleman, H., Roy, K.: Digital CMOS Logic Operation in the Sub-threshold Region. In: Proc. Tenth Great Lakes Symposium on VLSI, pp. 107–112. Chicago, IL (2000)
14. Soeleman, H., Roy, K., Paul, B.: Robust Subthreshold Logic for Ultra-low Power Operation. IEEE Transactions on Very Large Scale Integration (VLSI) Systems **9**(1), 90–99 (2001)
15. Soeleman, H., Roy, K., Paul, B.: Robust Subthreshold Logic for Ultra-low Power Operation. IEEE Transactions on Very Large Scale Integration (VLSI) Systems **9**(1), 90–99 (2001)
16. Wang, A., Chandrakasan, A., Kosonocky, S.: Optimal Supply and Threshold Scaling for Subthreshold CMOS Circuits. In: Proc. IEEE Computer Society Annual Symposium on VLSI, pp. 5–9 (2003)
17. Zhai, B., Blaauw, D., Sylvester, D., Flautner, K.: Theoretical and Practical Limits of Dynamic Voltage Scaling. In: Proc. Design Automation Conference, pp. 868–873. San Diego, CA (2004)

Chapter 12
Reclaiming the Sub-threshold Speed Penalty Through Micropipelining

12.1 Overview

Sub-threshold circuit design is an appealing means to dramatically reduce power consumption. However, sub-threshold designs suffer from the drawback of being significantly slower than traditional designs. To reduce the speed gap between sub-threshold and traditional designs, we propose a sub-threshold circuit design approach based on asynchronous micropipelining of a levelized network of PLAs. We describe the handshaking protocol, circuit design and logic synthesis issues in this context. Our preliminary results demonstrate that *by using our approach, a design can be sped up by about* 7×, with an area penalty of 47%. Further, our approach yields an energy improvement of about 4×, compared to a traditional network of PLA design. Our approach is quite general and can be applied to traditional circuits as well.

The key contribution of this work is to come up with a technique that enjoys an extreme low power consumption due to the use of sub-threshold circuitry, but at the same time compensates for the sub-threshold delay penalty. Such techniques would widen the applicability of sub-threshold circuit design approaches to a broader class of applications. *The proposed approach utilizes a network of PLA (NPLA) based sub-threshold circuit design approach, configured in an asynchronous micropipelined structure to enhance the speed of the circuit.* Sub-threshold circuit design has so far been used in only simple digital circuits and analog circuits. The design methodologies used in implementing such circuits are adhoc. *Our approach provides a systematic EDA framework for the design of complex digital systems using sub-threshold NPLA circuits.* It additionally utilizes an asynchronous micropipelining approach to speed up the sub-threshold design. Our experiments indicate that this approach yields a significant circuit speedup and improvement in energy consumption compared to traditional NPLA designs. Circuit speedup is measured in terms of computational throughput. In Sect. 12.2, we provide details about our micropipelined PLA-based asynchronous protocol and the logic synthesis approach to decompose a circuit into this circuit paradigm. The delay, area, power and energy characteristics of designs, which are implemented using our approach, are given in Sect. 12.3.

N. Jayakumar et al., *Minimizing and Exploiting Leakage in VLSI Design*,
DOI 10.1007/978-1-4419-0950-3_12, © Springer Science+Business Media, LLC 2010

12.2 Our Approach

Our approach to enhancing the speed of sub-threshold circuits is based on implementing the circuit using a micropipelined asynchronous network of PLAs. This implementation has the advantage of increasing the throughput of the circuit to a constant, regardless of the topological depth of the circuit. PLAs with adjacent topological depths in this structure communicate via an asynchronous handshake, which ensures correct operation of the design.

In Sect. 12.2.1, we describe the operation of the asynchronous micropipeline, along with its handshaking protocol. Section 12.2.2 indicates our approach for synthesizing a network of PLAs from a multi-level logic circuit, in a manner which is optimized for an asynchronous micropipeline-based implementation. We point out that in addition to PLAs, this methodology requires a specialized circuit block (which we call a *stutter* block), which delays signals that traverse multiple levels in the NPLA. Section 12.2.3 describes the design of a single PLA in this methodology and the handshaking logic within each PLA. We also discuss details of each PLA (maximum number of inputs, outputs and rows) used in our approach. We also describe the design of the stutter blocks used in our approach.

12.2.1 Asynchronous Micropipelined NPLAs

The concept of micropipelines was first introduced by Ivan Sutherland at his Turning Award lecture [7] in 1989. Our asynchronous micropipelined design methodology is based on the use of NPLAs [3, 4]. The choice of PLAs for the implementation of the underlying logic is that these structures can be designed to have a constant output delay across all possible input combinations. Also, the use of precharged NOR-NOR PLAs results in a compact and fast circuit. It was shown that for a single PLA, the delay was about 48% and the area about 46% compared to a standard cell based design [4], as long as the PLA was medium-sized (with 7–15 inputs, 5–10 outputs and 15–30 rows).

For a robust asynchronous micropipelined implementation, it is critical that the delays of the underlying circuit blocks are extremely predictable. The constant delay of a dynamic PLA over all input combinations makes it a very attractive choice in this context. Also, we utilize PLAs of fixed size in our approach. In this way we satisfy this important requirement of predictable delay. Note that in a sub-threshold design methodology, circuit delays vary significantly as a function of process, temperature and voltage (PVT) variations, as indicated in Chap. 9. However, we propose to use an on-the-fly, dynamically delay-compensated NPLA structure, which was shown (in Chap. 10) to dramatically reduce this variation. The residual variation in NPLA delay after applying this technique is minimal. Therefore, a simple guard-banding can achieve a predictable PLA delay across PVT variations in a sub-threshold context.

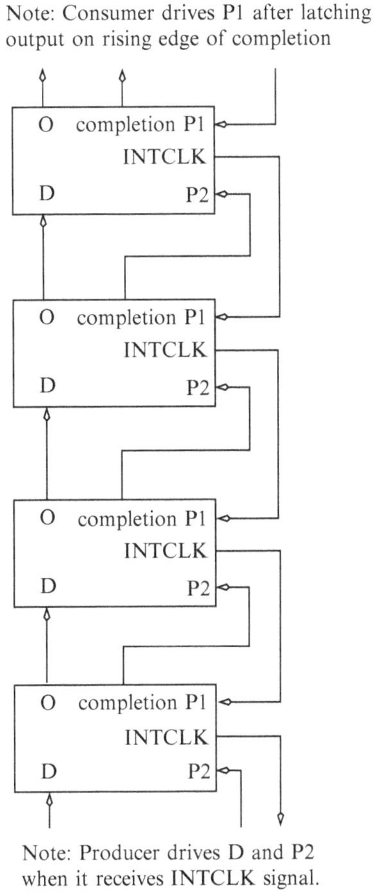

Fig. 12.1 NPLA-based asynchronous micropipelined circuit

The structure of the asynchronous micropipelined NPLA is shown in Fig. 12.1. Each PLA is a precharged NOR-NOR structure. However, the determination of when a PLA precharges and evaluates is made based on the handshaking protocol. There is no global clock signal in the design. Each PLA has a *completion* signal (which is assumed to switch *high* when evaluation of the PLA completes), which indicates that its outputs have been computed. In Fig. 12.1, the inputs of a PLA are indicated as D and the PLA outputs are marked as O. Each PLA has two inputs $P1$ and $P2$, which control the asynchronous handshake signal marked *completion* that indicates when the PLA has completed an evaluation or precharge operation. The *completion* signal of a PLA switches high when the PLA completes an evaluation operation and switches low when it completes a precharge operation. Each PLA also has an internally generated clock signal (marked *INTCLK*). The PLA precharges when *INTCLK* is low and evaluates when it is high.

The precharge operation of a PLA begins when P1 goes high, while evaluation starts when $P2$ rises, provided the *completion* signal of the PLA is low. After the *completion* signal of the topologically lowest level (level 1) PLAs goes low (PLA has precharged), the $P2$ signal of the topologically lowest level PLAs is asserted. This causes level 1 PLAs to evaluate. When the *completion* signal of the level 1 PLAs is asserted, the level 2 PLAs begin evaluation. When the level 2 PLAs start evaluating (a short period after the INTCLK signal of the level 2 PLA rises), the level 1 PLAs start precharging. This ensures that the data from the PLAs of level 1 to the PLAs of level 2 are held until the PLAs of level 2 have latched the data from PLAs of level 1. This is necessary to make sure that data are not lost in the micropipeline. This handshaking mechanism is utilized across all PLA levels. Its implementation is shown in Fig. 12.2.

The micropipelined structure in Fig. 12.1 shows a single PLA at any topological level. In practice, there may be several PLAs at any level, in which case the *completion* signal for any level i would be generated by logically ANDing the *completion* signals of all PLAs of level i.

The screen capture of a Verilog simulation for a series of four PLAs showing the working of our handshaking protocol is shown in Fig. 12.3. Note that this figure illustrates the asynchronous nature of the computation. In this figure, P2 is a signal from outside the micropipeline that signals the level 1 PLA to start evaluating (if the

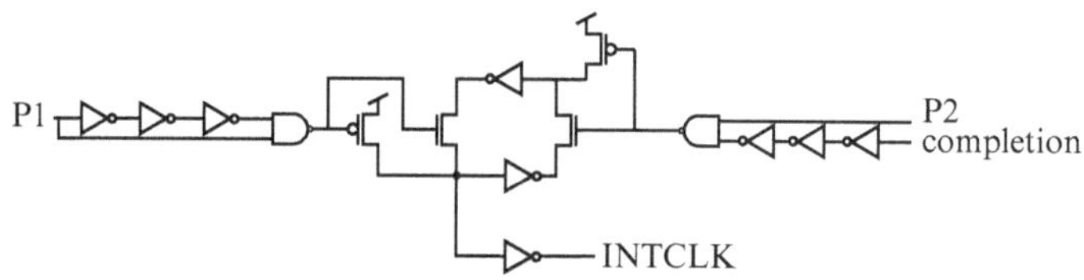

Fig. 12.2 Micropipelined PLA handshaking logic

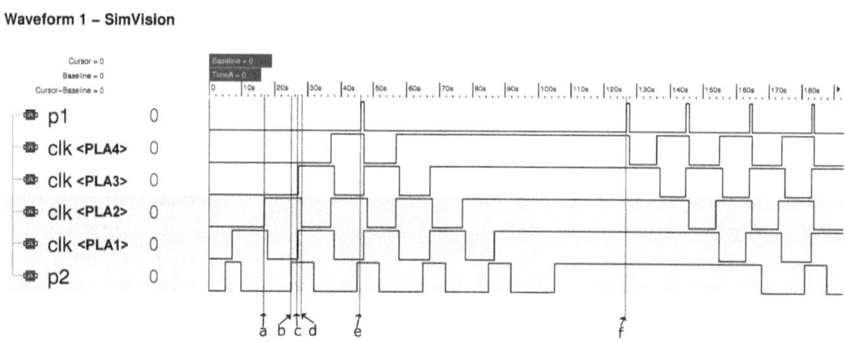

Fig. 12.3 Verilog simulation of our approach

level 1 PLA is precharged). Once the level 1 PLA completes evaluation it signals the level 2 PLA to start evaluating. This happens at the time instant marked a, which occurs at a short handshake period after the level 1 PLA completed its evaluation. We call this handshake period the *evaluation handshake period*. The level 2 PLA completes its evaluation at the time instant marked b and then after a period equal to the evaluation handshake period, the level 3 PLA starts evaluating at the time instant marked c. A short period after this (at the time instant marked c), the level 2 PLA starts precharging. We call this short period the *precharge handshake period*. P1 is the user acknowledgment signal generated (at time instant marked e) after the PLA at level 4 completes its evaluation and the user has latched the data from the PLAs at this level. When the level 4 PLA receives this signal it starts precharging. If the user is late in acknowledging the data from the PLA at the last level, the pipeline is stalled till P1 is asserted again (at time instant marked f).

12.2.2 Synthesis of Micropipelined PLA Networks

Synthesis of a PLA network for an asynchronous micropipelined implementation consists of a two-step process. In the first step, we generate a NPLA from a multi-level logic netlist. In the second, we infer the stuttered signals that are induced by the synthesized result and augment the netlist of the first part with stutter blocks, which delay signals that traverse more than one level of PLAs.

In the first step, we begin by performing technology-independent optimizations on the multi-level circuit C. Next, we decompose C into a network C^* of nodes with at most p inputs. In our experiments, $p = 5$. Now C^* is sorted in depth-first manner. The resulting array of nodes is sorted in *levelization*[1] order and placed into an array L.

Now we greedily construct the logic in each PLA, by successively grouping nodes from L such that the resulting PLA implementation of the grouped nodes N^* does not violate the constraints of PLA width and height. This check is performed in a *check_PLA* routine, which first flattens N^* into a two-level form, P. It then calls *espresso* [1] on the result to minimize the number of cubes in P. Next, *check_PLA* calls a *PLA folding* routine that attempts to fold the inputs of P so as to implement a more complex PLA in the same area. Finally *check_PLA* ensures that the final PLA, after folding and simplification using *espresso*, satisfies the maximum width and height constraints, respectively. If so, we attempt to include another node into N^*; otherwise we append the last PLA satisfying the height and width constraints to the result.

The *get_next_element* routine returns the most favorable node n among nodes in the fanout of nodes $n' \in N^*$ and nodes n'', which have the same level as the first

[1] Primary inputs are assigned a level 0, and other nodes are assigned a level that is one larger than the maximum level of all their fanins.

```
Algorithm Decompose_Circuit_to_NPLA
C = optimize_network(C)
C* = decompose_network(C, p)
L = dfs_and_levelize_nodes(C*)
N* = 0
RESULT = 0
while get_next_element(L) != NIL do
   N* = N* ∪ get_next_element(L)
   P = make_PLA(N*)
   if check_PLA(P, W, H) then
      continue
   else
      Q = remove_last_element(N*)
      RESULT = RESULT ∪ N*
      N* = Q
   end if
end while
```

Fig. 12.4 Decomposition of a circuit into a network of PLAs

node included into N^*, provided that the inclusion of n into N^* would not result in a cyclic PLA network . If such nodes are not available, the first unmapped node from L is returned. The favorability of a candidate is computed as:

$$favorability(n) = 2 \times [\#common\ fanins(n, n')] + [\#common\ fanouts(n, n')].$$

Nodes with shared fanins and fanouts decrease the number of PLAs created. We also found that shared fanins had a greater effect on this decrease. Hence, in evaluating the favorability of a node we gave a greater weight to those nodes that shared a fanin with a node already included in the current PLA.

We implemented the algorithm to decompose a circuit into a network of PLAs in SIS [6]. The pseudo-code of the algorithm is shown in Fig. 12.4.

The PLAs we used in our experiments had 16 inputs, 14 outputs and 24 rows. We found, through extensive experiments, that this size yielded a small number of PLAs and stutter blocks for a set of benchmark circuits.

Inferring of stuttered signals is performed by traversing the network of PLAs from inputs to outputs. For any output of a PLA of level l, if the PLAs in its fanout have a maximum level of l_j, then $l_j - l - 1$ stutter signals are inserted for this output, one for every level between l and l_j.

12.2.3 Circuit Details of PLAs and Stutter Blocks

The PLA we use is a precharged NOR-NOR PLA (similar to the ones used in Chaps. 10 and 11). The major difference between the PLAs utilized in this chapter and the ones utilized in Chaps. 10 and 11 is that the inputs have latches to store

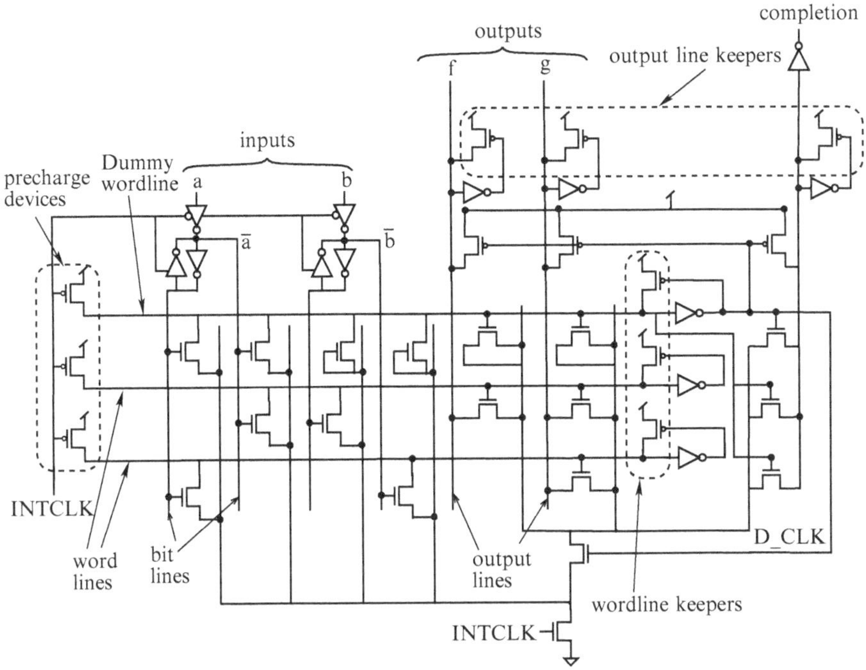

Fig. 12.5 Schematic of the PLA

the data from a previous level. The schematic view of the PLA circuit is shown in Fig. 12.5. The wordlines of the PLA (which represent the cubes of the function to be implemented) run horizontally through the AND and OR plane of the PLA. The bit lines (which carry the inputs and their complements) run vertically through the AND plane, while the output lines run vertically through the OR plane of the PLA structure. The layout view of our PLA is shown in Fig. 12.6. The operation of the PLA is similar to that of the non-micropipelined PLAs in Chaps. 10 and 11 (with *INTCLK* replacing *CLK*). The operation is explained here again for the reader's convenience.

INTCLK is an internal clock signal manipulated by the micropipelining protocol. When INTCLK (which is manipulated by the micropipelining handshake protocol) is low, the PLA enters the *precharge* phase. During this time, the horizontal wordlines get precharged. A special wordline (the dummy wordline), which is the maximally loaded wordline, also gets precharged. This forces the signal D_CLK to go low, cutting off the OR plane from GND and causing the output lines to also get precharged. A special output line (which is inverted to produce the signal *completion* shown in Fig. 12.5) also gets precharged. The dummy wordline is designed to be the last wordline to switch (by making it maximally loaded among all wordlines). Similarly, the *completion* line is also the last output line to switch, since it is maximally loaded as well, in comparison to other outputs. The *completion* line

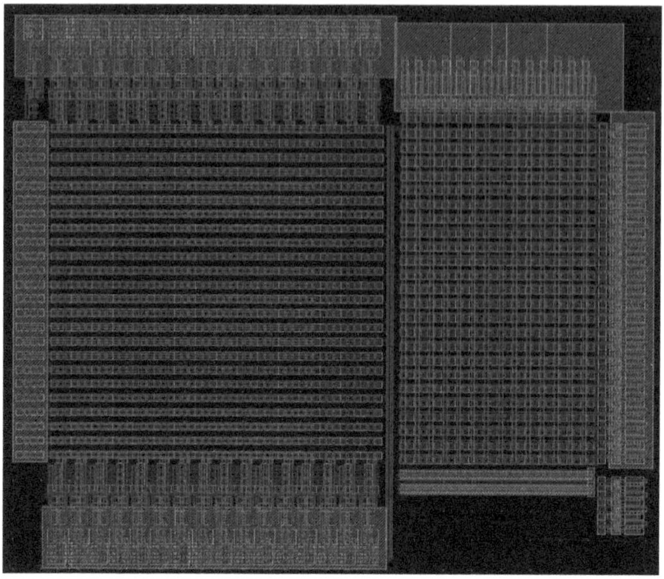

Fig. 12.6 Layout view of the PLA

switching low signals the completion of the precharge operation of the PLA. In the precharged state, all the wordlines and the output lines of the PLA are precharged. Now, when INTCLK switches high, the PLA enters the *evaluation* phase. In evaluation, if any of the vertical bitlines are high, the wordline that it is connected to gets pulled low. One of the inputs *and* its complement are connected to the dummy wordline, so that the dummy wordline switches low during *every* evaluate phase. By design, the dummy wordline is the last wordline to switch low. This makes the signal D_CLK go high, as a result of which the GND gating transistor in the OR plane now turns on.[2] The output lines to which wordlines that have switched low are connected will switch low. The *completion* line which is connected to the complement of the dummy wordline is the last line to switch high. This signals the completion of the evaluation operation.

The INTCLK signal is generated from the *completion*, $P1$ and $P2$ signals using the circuit shown in Fig. 12.2. On every rising edge of $P1$, a pulse is generated, which makes the INTCLK signal go low, forcing the PLA to enter the precharge phase. In other words, PLA p enters the precharge phase if PLAs at a level above the PLA p have started evaluation (after latching the input data). Once this happens, the *completion* signal of the PLA p falls (after all other signals in p have precharged). At this point, if $P2$ rises, then the PLA p enters the evaluation phase. In other words, if the PLA p has been precharged, and if the PLAs a level below

[2] Note that in the sub-threshold region a transistor is either *off* or *less off*. For the sake of simplicity, we say that an NMOS transistor is *on* when its gate is at VDD and *off* when its gate is at GND. Similarly we say a PMOS transistor is *on* when its gate is at GND and *off* when its gate is at VDD.

complete their computation, then p enters the evaluation phase. The additional inverter(s) in the path of the *completion* signal are for design guard-banding. In our SPICE [5] simulation of this handshaking block, we found that it had a worst case delay of 25 ns for INTCLK to fall, measured with respect to $P1$ rising. We called this the *precharge handshake period* in Sect. 12.2.1. The handshaking block had a worst case delay of 60 ns for INTCLK to rise (measured with respect to *completion* falling). We called this the *evaluation handshake period* in Sect. 12.2.1.

Note that each of the PLAs has a set of level-sensitive latches on its inputs. When the PLA p has completed its computation, these latches hold their state, ensuring that the precharging of PLAs a level below does not change the state of the outputs of p that have been computed.

In this manner, odd levels of the NPLA precharge while even levels of PLAs evaluate.

The stutter block is simply a series of latches, implemented in the footprint of a PLA (in terms of height). Its function is to delay signals that traverse across levels of PLAs, in order to guarantee correct operation under asynchronous micropipelining. For example, if there is a signal S_{jump1} that is an output of a level 1 PLA and is an input to a level 3 PLA, then a stutter block, consisting of a single latch, is placed between the two PLAs. The signal S_{jump1} is used as the data input to this latch and the data are latched using the INTCLK signal from level 2 PLA(s). This ensures that all the inputs to the level 3 PLA(s) are ready at the same time. For a signal traversing across n levels, n latches are required.

12.3 Experimental Results

To compare the characteristics of an asynchronous micropipelined network of PLAs with that of a network of PLAs, we performed extensive simulations. All circuit simulations were done in SPICE [5], assuming a supply voltage of 0.2 V and a temperature of 25°C and using 65 nm BPTM [2] model cards. The area of the two design styles was computed using the sum of the areas of all the PLAs in the design, including the area of any stutter blocks (in the case of the micropipelined network of PLAs).

The asynchronous micropipelined network of PLAs has a throughput of

$$T = \frac{1}{T_{\text{eval}} + T_{\text{pchg}} + 2 \cdot H_{\text{eval}} + H_{\text{pchg}}}.$$

Here T_{eval} is the evaluation delay of the PLA (recall, we utilize fixed sized PLAs in the design), T_{pchg} is the precharge delay of the PLA, H_{eval} is the evaluation handshake period and H_{pchg} is the precharge handshake period. The values of T_{eval}, T_{pchg}, H_{eval}, H_{pchg} are 210 ns, 155 ns, 60 ns and 25 ns, respectively. As a consequence, the throughput is $\frac{1}{510\,\text{ns}}$. Note that the latency is still proportional to the number of PLA levels in the design, but the throughput is a constant.

In the traditional network of PLA implementation, all levels of PLAs are precharged together and then evaluate in a domino fashion. The timing diagram of this is shown in Fig. 11.7 in Chap. 11. In case of the traditional network of PLA implementation, the delay is given by the topological depth of the PLA network (in terms of number of PLAs) times the evaluation delay T_{eval} of each PLA. We also add to this the time taken to precharge all the PLAs in the design. Note that, in general, this is substantially greater than the throughput of our micropipelined approach.

We also compared the energy consumption of the two types of implementations. More specifically we compared the energy consumption per computation in the two types of NPLAs. For the micropipelined implementation, we first found (through SPICE simulation) the energy consumption for the operation of 1 PLA (over a period of 510 ns) and multiplied this by the number of PLAs. To this, we add the energy consumption of the handshaking logic and the energy consumption in the stutter blocks. This gives us the energy consumption for one computation through the micropipelined NPLA.

While a micropipelined PLA spends very little time (equal to the handshaking periods) in a *precharged* state or *evaluated* state, the traditional NPLA spends substantial periods of time in the *precharged* state and *evaluated* state. This is evident from the timing diagram shown in Fig. 11.7. As a consequence, the micropipelined network of PLA-based design wastes less energy in leakage than traditional network of PLA-based designs.

Table 12.1 reports the results of our experiments. The first column represents the circuit under study. The second column reports the number of PLAs required, while the third column reports the number of stutter blocks in the micropipelined network of PLAs. The next three columns report the delay of the non-micropipelined PLA, the throughput of the micropipelined PLA, and their ratio. Note that the throughput of the micropipelined PLAs is constant. The traditional PLA network delay is computed as described above. We note that the *micropipelined PLA results in a speedup of about 7× over a traditional design*. This is because in the micropipelined network of PLA circuit, the measure of delay is its throughput. Hence, for the network of PLA circuits with larger topological depths, this improvement is more pronounced. Columns 7, 8 and 9 indicate that the energy consumption of the micropipelined NPLAs is about 4× lower than the energy consumption of the traditional NPLAs. The area penalty for the approach is about 47% on average, as indicated in the last three columns of Table 12.1.

12.4 Optimum VDD for Micropipelined NPLAs

In the previous chapter (Chap. 11), we discussed how the optimum supply voltage (VDD) that minimizes energy consumption for Network of PLAs depends on the logic depth of the network. The optimum VDD is higher for a circuit with a larger logic depth. This is due to the fact that while one PLA is precharging or evaluating,

Table 12.1 Comparison of micropipelined with traditional circuits

Ckt	No. of PLAs	No. of Stutter blocks	Delay (ns) ↓			Energy (fJ) ↓			Area (μ^2) ↑		
			Non-μpipe	μpipe	Impr.	Non-μpipe	μpipe	Impr.	Non-μpipe	μpipe	Ovh
alu4	14	5	2,885	510	5.66	5,984.80	1,811.43	3.30	9,408	12,768	1.36
apex6	24	12	2,465	510	4.83	9,033.09	3,261.19	2.77	16,128	24,192	1.50
C432	11	4	2,255	510	4.42	3,877.22	1,397.00	2.78	7,392	10,080	1.36
C499	14	4	2,255	510	4.42	4,961.02	1,768.64	2.80	9,408	12,096	1.29
C880	16	5	2,255	510	4.42	6,088.11	2,052.22	2.97	10,752	14,112	1.31
C1355	21	10	3,305	510	6.48	10,198.86	2,863.68	3.56	14,112	20,832	1.48
C1908	24	13	3,935	510	7.72	13,814.19	3,307.96	4.18	16,128	24,864	1.54
C2670	34	13	3,515	510	6.89	18,694.33	4,472.11	4.18	22,848	31,584	1.38
C3540	67	46	7,505	510	14.72	73,900.56	9,777.18	7.56	45,024	75,936	1.69
pair	65	35	4,565	510	8.95	44,442.77	9,047.27	4.91	43,680	67,200	1.54
rot	19	13	3,095	510	6.07	8,966.68	2,774.15	3.23	12,768	21,504	1.68
Avg	28.09	14.55			6.78			3.84			1.47

Table 12.2 Optimum VDD shift with PLA size

Size of PLA			Optimum VDD (V)			
No. of inputs	No. of outputs	No. of rows	At 25°C	At 50°C	At 75°C	At 100°C
16	14	24	0.22	0.28	0.30	0.30
16	10	16	0.22	0.28	0.30	0.30
12	6	12	0.20	0.28	0.28	0.30
8	4	8	0.18	0.22	0.22	0.28
4	2	4	0.15	0.18	0.20	0.22

the other PLAs in the circuit waste energy in the idle *precharged* and *evaluated* states. In a micropipelined PLA, very little time is spent in these idle states. Hence, the optimum VDD is expected to be low. The energy consumed by each PLA in a micropipelined Network of PLAs is equal to the sum of the energies spent in the *evaluating* and *precharging* states and the energies spent in the *precharged* states and *evaluated* states during the handshake periods. For our micropipeline, we hence estimate the energy consumed by each PLA to be given by the following formula

$$\begin{aligned}\text{Energy} &= PchgEnergy_{dyn} + EvalEnergy_{dyn} \\ &+ [PchgPwr_{sta} \times (H_{\text{eval}})] \\ &+ [EvalPwr_{sta} \times (H_{\text{eval}} + H_{\text{pchg}})].\end{aligned} \quad (12.1)$$

We characterized PLAs of different sizes to explore how the size of the PLA would affect the optimum VDD point. The results are given in Table 12.2. The PLAs were characterized using SPICE and the energy estimated using (12.1).

As the data in Table 12.2 show, the optimum VDD is low since the PLAs spend very little time in the *precharged* and *evaluated* states. However, we do notice that as the PLA gets smaller, the optimum VDD does reduce. Also, just like we saw in the previous chapter, a higher temperature shifts the optimum VDD to higher value.

12.5 Summary

In recent times, power consumption has become a dominant issue in VLSI circuit design. Sub-threshold circuit design is an appealing means to dramatically reduce this power consumption. However, sub-threshold designs suffer from the drawback of being significantly slower than traditional designs. In this chapter, we described a means to reclaim the speed penalty associated with sub-threshold designs. The approach is based on the use of a sub-threshold circuit design approach, which is based on asynchronous micropipelining of a levelized network of PLAs. We have developed a handshaking protocol, a circuit design approach and logic synthesis methodologies in this context. Our preliminary results demonstrate that by using our approach, a design can be sped up by 7×, with an area penalty of 47%. Further, the energy consumption of micropipelined NPLA-based circuits is about 4× lower than that of the traditional NPLAs circuits. Our simulations were validated in VERILOG, and circuit level characteristics were extracted using SPICE modeling. Using the techniques described in Chap. 11, we also found that the optimal VDD for minimum energy operation of a micropipelined Network of PLAs can be above V_T (depending on the size of the PLA and the operating conditions). The techniques described in this chapter are equally applicable for these operating conditions as well.

References

1. Brayton, R.K., Hachtel, G.D., McMullen, C.T., Sangiovanni-Vincentelli, A.: Logic Minimization Algorithms for VLSI Synthesis. Kluwer Academic Publishers, New York, NY (1984)
2. Cao, Y., Sato, T., Sylvester, D., Orshansky, M., Hu, C.: New Paradigm of Predictive MOSFET and Interconnect Modeling for Early Circuit Design. In: Proc. IEEE Custom Integrated Circuit Conference, pp. 201–204. Orlando, FL (2000). http://www-device.eecs.berkeley.edu/ ptm
3. Khatri, S.P.: Cross-talk Noise Immune VLSI Design Using Regular Layout Fabrics. Ph.D. thesis, University of California, Berkeley (1999)
4. Khatri, S.P., Brayton, R.K., Sangiovanni-Vincentelli, A.: Cross-talk Immune VLSI Design Using a Network of PLAs Embedded in a Regular Layout Fabric. In: Proc. IEEE/ACM International Conference on Computer Aided Design, pp. 412–418. San Jose, CA (2000)
5. Nagel, L.: SPICE: A Computer Program to Simulate Computer Circuits. In: University of California, Berkeley UCB/ERL Memo M520 (1995)
6. Sentovich, E.M., Singh, K.J., Lavagno, L., Moon, C., Murgai, R., Saldanha, A., Savoj, H., Stephan, P.R., Brayton, R.K., Sangiovanni-Vincentelli, A.L.: SIS: A System for Sequential Circuit Synthesis. Tech. Rep. UCB/ERL M92/41, erl, University of California, Berkeley, CA 94720 (1992)
7. Sutherland, I.E.: Micropipelines. Communications of the ACM **32**(6), 720–738 (1989)

Chapter 13
Part II: Conclusions and Future Directions

While the first part of this book discussed leakage reduction techniques, the second part focused on leakage *exploitation*. In Chap. 9 we first presented data from some exploratory studies that revealed the opportunity that sub-threshold circuit design offers. The main advantages of sub-threshold circuits are as follows:

- Low power consumption and heat dissipation
- Smaller delays with increasing temperature
- High power-delay product (PDP)

We also presented the three main disadvantages facing sub-threshold circuit design today:

- Large delay
- Sensitivity to process, voltage and temperature (PVT) variations
- Lack of a systematic EDA framework to implement sub-threshold circuits.

This chapter also discussed the application space for sub-threshold design. The remaining chapters of Part II of this book proposed techniques to address each of the disadvantages cited above.

In Chap. 10 we presented a way to make a sub-threshold circuit less sensitive to PVT variations. We proposed a sub-threshold design approach, which dynamically compensates for inter and intra-die PVT variations. The approach we proposed involved adaptively adjusting the body bias to dynamically stabilize the delay of the circuit. In the proposed approach a multi-level network of medium-sized Programmable Logic Arrays (PLAs) was the circuit implementation structure. The approach used a global *beat clock* and attempted to "phase lock" the delay of a representative PLA (in a cluster of localized PLAs) to the beat clock. This phase locking was done in a closed-loop fashion using a phase detector and charge pump, which charged or discharged the bulk node of the NMOS devices in the PLAs. The PLAs we used were dynamic (NOR-NOR) PLAs. In such PLAs, the critical delay (the evaluation delay) is dependent mainly on the NMOS devices in the core of the PLA. Hence, we only controlled the bulk nodes of the NMOS devices. Simulation results (using 65-nm BSIM4 model cards from [1]) proved that our adaptive body biasing scheme is very effective. An analysis of the loop gain of the closed-loop adaptive body biasing scheme was also presented. We found that the width of the

N. Jayakumar et al., *Minimizing and Exploiting Leakage in VLSI Design*, DOI 10.1007/978-1-4419-0950-3_13, © Springer Science+Business Media, LLC 2010 charge-pump transistors and the capacitance of the bulk node can be used to tune the response of the scheme. Sub-threshold circuits are extremely sensitive to PVT variations. A compensating scheme such as the one presented in Chap. 10 is crucial for any practical sub-threshold design.

While a lower voltage reduces power consumption it also worsens the time taken to perform a computation. As a result the energy consumed in performing a computation can actually be higher for a circuit utilizing a lower operating voltage. The optimum voltage for minimum energy is in fact dependent on the circuit topology. In Chap. 11, we studied the problem of finding the optimum voltage for minimum energy in the context of designing a circuit using a network of dynamic NOR-NOR PLAs. We derived a method to calculate the energy consumed by a network of medium (fixed) sized PLAs by just characterizing *one* of the PLAs in the network. Using this method we estimated the energy for networks of PLAs of various logic depths. We found that as the logic depth of a circuit got larger, the optimum VDD became higher. This is because when one PLA in a network is evaluating or precharging, the other PLAs in the network (at a different logic depth) are in the evaluated or precharged idle states, wasting leakage power. The dependence of the optimum VDD on circuit topology holds for other circuit design styles as well, not just for a network of PLA-based design.

In Chap. 12 we proposed using asynchronous micropipelining to help improve the throughput of sub-threshold circuits and hence reduce the speed gap between sub-threshold and traditional circuits. The approach used a network of PLA-based design flow similar to the flow used in Chaps. 10 and 11. The synthesis algorithm used in the design flow was augmented to allow the network of PLAs to be micropipelined. On a set of benchmark circuits, the micropipelined approach was found to give a 7× improvement in throughput over a non-micropipelined network of PLAs. After applying the micropipelining approach, the delay of a sub-threshold circuit is approximately 1.5–4× worse than a traditional super-threshold circuit. Without this technique, recall that the delay penalty was 10–25×. The micropipelined circuits were also found to be more energy efficient due to the fact that little time and energy was wasted in the idle precharged and evaluated states. Using the concepts of Chap. 11, we studied how the optimum VDD for an asynchronous micropipelined circuit would change with PLA size and temperature. We found that in a majority of cases, the optimum VDD for minimum energy was slightly above the threshold voltage of the NMOS devices. The micropipelining technique is applicable in these near-threshold regions of operation as well.

In Chaps. 10–12 we proposed using a network of PLAs to design sub-threshold circuits. In Chaps. 10 and 12 we presented approaches to respectively tackle the issues of sensitivity of sub-threshold circuits to PVT variations and the problem of increased delay of sub-threshold circuits. We also proposed design flows to implement digital circuits as a sub-threshold network of PLAs. As discussed in [2], using a network of medium-sized PLAs is a suitable way to implement structured ASICs with a low NRE. Structured ASICs allow designs to be implemented using very few lithography masks (metal and via masks only in the case of [2]). The sub-threshold design approaches presented here are, hence, very easily applied to a structured ASIC setting as well.

In a sub-threshold design, a high-quality power and ground distribution network is crucial since the operating voltages are extremely low. Also, such a circuit can be susceptible to noise. In such a scenario, a layout fabric [3, 4] is ideally suited for sub-threshold circuits. The network of PLAs used in our sub-threshold circuit design flows is naturally amenable to such a fabric. One of the reasons for the success of traditional standard-cell-based CMOS design technology is the existence of a design flow and methodology that made the design of standard-cell based ICs practical and feasible. The sub-threshold design approaches in this part of the book are presented to provide a design flow and methodology that can help make sub-threshold circuit design practical and feasible.

Sub-threshold circuits are useful in applications where minimum power and energy consumption are most important while performance is a secondary requirement. Examples of such applications are sensor networks, digital wrist watches and medical equipment such as hearing aids. Another possible application for sub-threshold circuits is the following – in the near future, we could have devices implanted within our bodies, which monitor the status of our health. These devices could probably derive their energy from the heat in the body or the flow of blood. These devices will be required to consume and dissipate extremely low amounts of power not only because the energy available is limited, but also because the heat dissipated by the device should not affect the surrounding tissue that it is implanted in. In such applications, sub-threshold designs are probably going to be the only feasible choice. With a large market for such low power devices, sub-threshold circuit design could become as popular as traditional CMOS design. The sub-threshold design approaches presented in this book should help accelerate the adoption of such devices.

References

1. Cao, Y., Sato, T., Sylvester, D., Orshansky, M., Hu, C.: New Paradigm of Predictive MOSFET and Interconnect Modeling for Early Circuit Design. In: Proc. IEEE Custom Integrated Circuit Conference, pp. 201–204. Orlando, FL (2000). http://www-device.eecs.berkeley.edu/ ptm
2. Jayakumar, N., Khatri, S.: A METAL and VIA Maskset Programmable VLSI Design Methodology Using PLAs. In: Proc. IEEE/ACM International Conference on Computer Aided Design, pp. 590–594. San Jose, CA (2004)
3. Khatri, S., Mehrotra, A., Brayton, R., Sangiovanni-Vincentelli, A., Otten, R.: A Novel VLSI Layout Fabric for Deep Sub-Micron Applications. In: Proc. Design Automation Conference. New Orleans, LA (1999)
4. Khatri, S.P., Brayton, R.K., Sangiovanni-Vincentelli, A.: Cross-talk Immune VLSI Design Using a Network of PLAs Embedded in a Regular Layout Fabric. In: Proc. IEEE/ACM International Conference on Computer Aided Design, pp. 412–418. San Jose, CA (2000)

Part III
Design of a Sub-threshold BFSK Transmitter IC

In the first part of this book, techniques to minimize leakage were presented. In the second part of the book, we presented sub-threshold circuit design methodologies. In the third part of this book, we present details of how we implemented and tested a robust sub-threshold design flow (which uses circuit level PVT compensation ideas from the second part of the book) to stabilize circuit performance.

We design and fabricate a sub-threshold wireless BFSK transmitter chip. The transmitter is specified to transmit baseband signals up to a data rate of 32 kbps over a distance of 1,000 m. In addition to the sub-threshold implementation, we implement the BFSK transmitter using a standard cell methodology on the same die operating at super-threshold voltages on a different voltage domain. Experiments using the fabricated die show that the sub-threshold circuit consumes *19.4*× lower power than the traditional standard cell-based implementation.

Outline of Part III

The main objective of this part of the book is to demonstrate the viability of a sub-threshold circuit design approach for use in designs that demand extreme low power consumption. There are currently no validated design flows or proven design methodologies for designing sub-threshold circuits. This part of the book attempts to do the following:

- To validate the sub-threshold circuit design techniques introduced in the second part of the book
- To come up with a robust design methodology to design and fabricate sub-threshold circuits
- To choose an application that will demonstrate the usefulness of a low power sub-threshold circuit
- To design the required circuit, fabricate and test the chip

- To quantitatively compare the post-silicon power consumption of a sub-threshold circuit implementation with that of a traditional standard cell-based implementation of the same circuit

In Chap. 14, we choose a test application that we will implement using sub-threshold circuits. We present a system level architecture and describe each of the system level blocks in detail. We then discuss the various design constraints and optimizations needed for the particular application. We then come up with a design framework to implement the design.

In Chap. 15, we present a detailed account of the steps involved in the implementation of the design. We explain the design flow used to implement the sub-threshold circuit, and we explain the circuit design of the required components. We also list out the validation methodologies used to verify the design before tapeout. We discuss several special features added to the design that facilitate debugging and testing. We also summarize some of the fail-safe mechanisms added to the design, which enable proper functionality even if some of the components failed to work as expected.

In Chap. 16, we quantitatively list the experiments performed and the results obtained from the fabricated die. We also show that the sub-threshold circuit consumes 19.4× lesser power than the standard cell circuit implementing the same function, under the specified operating conditions.

Chapter 14
Design of the Chip

14.1 Overview

This chapter presents the design of a test application that will utilize the circuit design methodologies described in Part II of this book. Sect. 14.2 discusses the criteria used to choose a test application and also an overview of what basic building blocks are required for such an application. It also defines the design constraints that are to be taken into account while designing a sub-threshold circuit. The architecture of the whole system and the details of the sub-blocks of the system are covered in Sect. 14.3. This chapter also outlines some special considerations and redundant features and failure-safe features that are built into the chip. The design of the chip is targeted for the TSMC [2] 0.25 μm process, which is a triple well CMOS process.

14.2 Test Vehicle

There is a large and growing application space that requires a very low power consumption without the need for high speed. One such application is a wireless radio transmitter, where the signal to be transmitted occupies a small bandwidth (such as voice). An ultra-low power implementation of a radio transmitter will have broad implications for the class of applications that demand very low power consumption. For example, this wireless transmitter can be used in sensor networks. In this design, the radio transmitter is realized with digital circuits as far as possible, since digital circuits are preferable to analog circuits when operating in the sub-threshold region. The digital circuits are implemented using a Network of PLA (NPLA) based approach. The immunity of the circuit to variation can be strengthened using the dynamic delay compensation circuitry that was introduced in Chap. 10) of this book. Details on the implementation of this compensation scheme in this design are presented in Sect. 14.3.3

We chose a simple digital modulation scheme for the radio transmitter. Binary Frequency Shift Keying (BFSK) and Binary Phase Shift Keying (BPSK) are two well-known digital modulation schemes. BPSK is 3 dB more power efficient than

N. Jayakumar et al., *Minimizing and Exploiting Leakage in VLSI Design*,
DOI 10.1007/978-1-4419-0950-3_14, © Springer Science+Business Media, LLC 2010

BFSK. However, BFSK has the advantage of being easy to implement. Hence, BFSK is used as the modulation scheme for our radio transmitter. The architecture of the system to be implemented is shown and the various sub-blocks in the system are explained in a detailed fashion in Sect. 14.2.1.

14.2.1 BFSK Radio Transmitter Architecture

A typical BFSK transmitter generates a frequency tone at the output and shifts the frequency of the output tone to pre-determined values depending on the value of the input, which can be a logical HIGH or LOW. A generic digital BFSK transmitter block diagram is shown in Fig. 14.1. The input to the transmitter is assumed to be digitized and supplied to the transmitter at a rate of R_B bits/s. The frequencies of the two tones that will be produced by the BFSK transmitter are given by f_1 and f_2. ϕ_1 and ϕ_2 are phase offsets that the two tones could have. Depending on the value of the binary input, one of the tones is multiplexed to the output. A BFSK transmitter can be coherent or non-coherent. In a coherent BFSK modulation scheme, $\phi_1 = \phi_2$ and in a non-coherent BFSK modulation scheme, $\phi_1 \neq \phi_2$. In practice coherent BFSK modulation is extremely hard to demodulate since there is a synchronization required between the transmitter and the receiver. Hence we use a non-coherent modulation scheme. For non-coherent modulation, if the BFSK modulation has the condition that $f_1 - f_2$ is an integer multiple of the input bit rate, R_B then the modulation is called orthogonal FSK (since the two signals used for modulating the binary data are orthogonal if this condition is met). If this condition is not met the FSK scheme is called non-orthogonal. The difference between the two schemes is that, non-orthogonal FSK requires more transmit power than orthogonal FSK for the same error performance at the receiver side. The receiver for an both schemes can be constructed using a couple of bandpass filters with their pass band frequencies centered around f_1 and f_2, respectively.

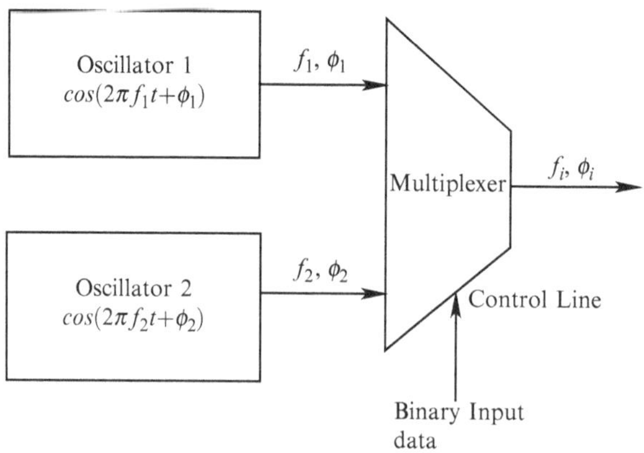

Fig. 14.1 BFSK transmitter architecture

While designing a BFSK transmitter, the two oscillators in Fig. 14.1 can be realized using digital circuits as a Numerically Controlled Oscillator (NCO), which will be described in Sect. 14.3.4.1. In order to do wireless transmission of a signal, we need a Digital to Analog Converter (DAC) and an antenna.

14.3 System Architecture

The BFSK transmitter architecture consists of a digital BFSK modulation circuit, a DAC, an amplifier and an antenna for wireless transmission. This is shown in Fig. 14.2. The BFSK modulator is implemented as a digital circuit, using a network of Programmable Logic Arrays (NPLAs). We first give a brief introduction to PLAs and how they are used in a network to do computations. The reader may skip this portion if he/she has already read about this in Part II of this book. We will also discuss in detail about each of the digital and analog components that make up the design of the system.

14.3.1 PLA Basics

This section describes the structure and operation of PLAs, which are the basic circuit modules used in this design. Note that the PLAs in this design operate in their sub-threshold region of conduction. The way in which logic is implemented in a PLA is discussed in Sect. 10.3.1.

The schematic of the PLA used in this design is shown in Fig. 14.3.

All the PLAs in our design are of the precharged NOR NOR type and have a fixed number of inputs (8), outputs (6) and cubes (12). This was found to be a good size for the design based on logic synthesis results explained in Sect. 15.3 while using medium-sized PLAs (5–15 inputs, 3–8 outputs and 10–20 rows). Initial simulations

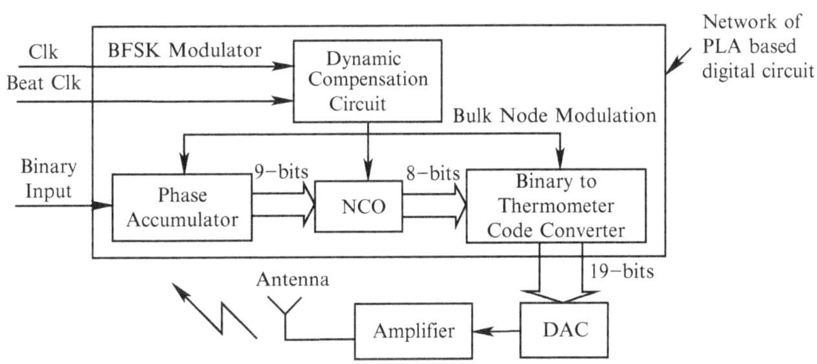

Fig. 14.2 System architecture

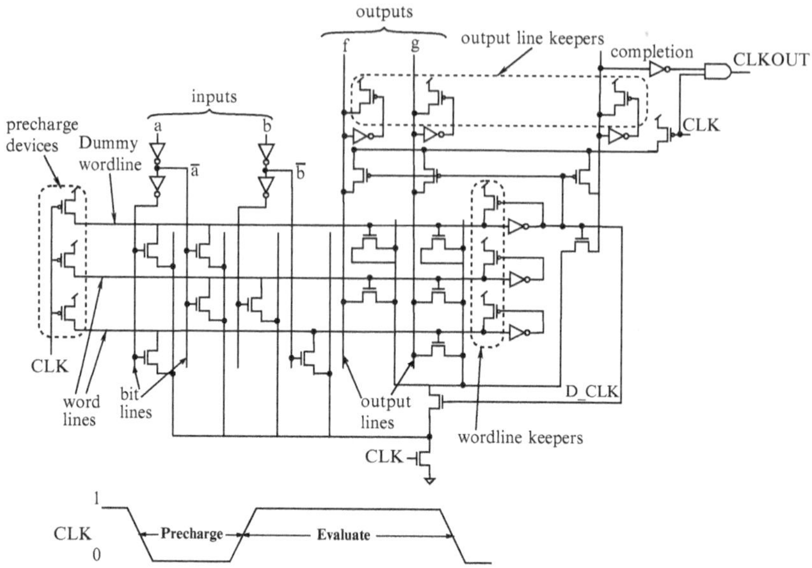

Fig. 14.3 Schematic view of PLA

using HSPICE [1] showed that precharge and evaluate time for the 8 input, 6 output, 12 cube NOR NOR PLA were $T_{\text{pchg}} = 45\,ns$ and $T_{\text{eval}} = 35\,ns$.

Also, a technique called folding is used to enhance a PLA to hold more logic without increasing the area used. This is done by running two unconnected *bit-lines* corresponding to two different inputs on the same track. One of the *bit-lines* start from the top of the PLA and the other one starts from the bottom and stops clear of the first *bit-line*. In this way, more cubes can be fitted into the PLA in a compact way.

14.3.2 Network of PLA Operation

A network of PLAs, NPLA is nothing but a multilevel network of PLAs. Each of the digital components that make up the digital BFSK modulator in Fig. 14.2, i.e. the Dynamic Compensation circuit, NCO and the Binary to Thermometer Code Converter are made of NPLAs. Each of these blocks are implemented as combinational circuits and the outputs of each of these blocks are registered using negative edge triggered flip-flops clocked by *Clk*. The flip-flops are negative edge triggered as the outputs of the flip-flops need to be stable when the *Clk* signal is HIGH when the PLAs are evaluating. The timing diagram of NPLAs in a single combinational circuit is shown in Fig. 14.4. Notice from this figure that all the PLAs in a network precharge at the same time and start evaluating one after another in a cascading fashion. Hence, an evaluation period has to be provided that is sufficient for *all* the PLAs to evaluate. Each PLA in the network is clocked by the previous PLAs *CLKOUT* signal except for the first PLA in the chain, which is clocked by the *CLK* signal. The

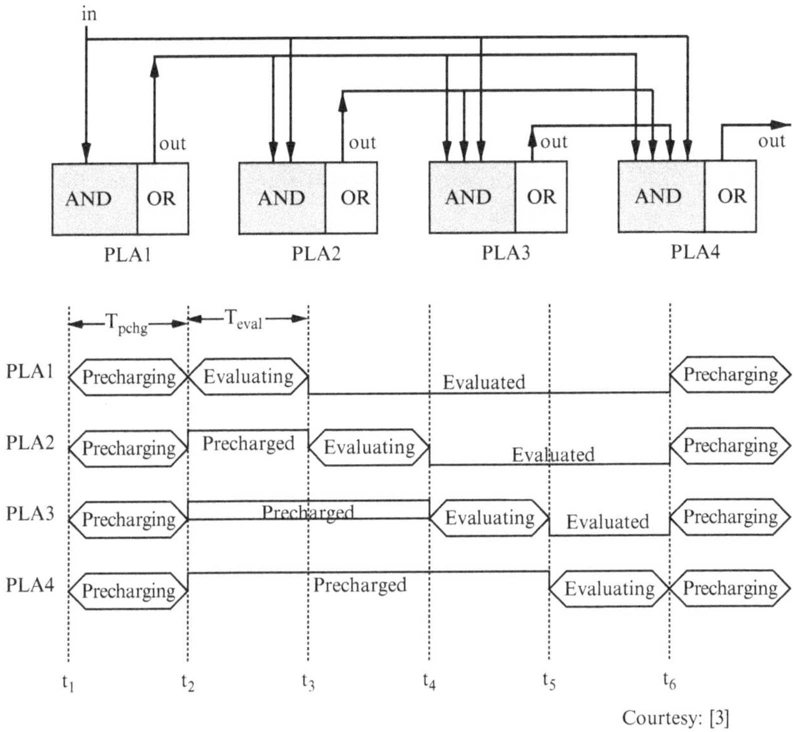

Fig. 14.4 Timing diagram of NPLAs

CLKOUT signal of each PLA is the logical AND of its *completion* signal and the *CLK* signal. The maximum throughput that can be achieved depends on the delay of the slowest combinational block. When implemented as a network of PLAs, the throughput of the circuit can be approximately written as:

$$\text{Throughput} = \frac{1}{T_{\text{pchg}} + N \times T_{\text{eval}}}. \tag{14.1}$$

Here N is the number of levels of PLAs needed in the multilevel network of PLAs. We will see in Sect. 15.3 that the maximum number of levels needed for the slowest combinational block for this design is 19. This gives us an estimate of the throughput as approximately 1.4 MHz, if we use $T_{\text{pchg}} = 45\,ns$ and $T_{\text{eval}} = 35\,ns$ as mentioned in the previous section.

14.3.3 Dynamic Compensation Circuit

As discussed in Sect. 10.4, the dynamic delay compensation circuit is used to to *phase lock* the circuit delay to a *beat clock*. The circuit in the design consists of a multi-level network of interconnected dynamic NOR-NOR PLAs. The total number of PLAs that are needed for this design is 33 as seen from Sect. 15.3. These PLAs are placed such that they are part of a single cluster of PLAs sharing a common Nbulk node. This Nbulk node is driven by a bulk bias adjustment circuit, which synchronizes the delay of a representative PLA in the cluster to a globally distributed *beat clock (BCLK)*. The beat clock is an external signal derived from the system clock.

The phase detector and charge-pump circuits used for the design are shown in Fig. 10.3.

The *BCLK* is used to speed up the operation of the PLAs during the evaluation phase. The evaluation delays of PLAs in our design happen one after the other as shown in Fig. 14.4. We need to choose a reference PLA out of the chain of PLAs in the network. The *completion* signal of this reference PLA is used as the reference circuit delay for the delay compensation circuit. Usually there are many levels of PLAs in the synthesized network of PLAs. In this scenario, it would be ideal to choose a PLA that completes its evaluation at approximately half the time it takes for the entire network of PLAs to complete its evaluation period. This is because the completion signal of the reference PLA would transition to a LOW value during the middle of the evaluation time span of the *CLK* signal. This gives the *BCLK* signal sufficient room on both sides of the completion signal to be able to generate equally long *pull-up* or *pull-down* signals. In our case, we use a PLA at logical depth 10 out of a maximum of 19 as the reference PLA.

14.3.4 The Digital BFSK Modulator

The function of the digital BFSK modulator as seen in Sect. 14.2.1 is to produce either of two frequency tones depending on the logical value of a binary input signal. The digital BFSK Modulator seen in Fig. 14.1 has two oscillators, but we have reduced this complexity of having two oscillators by using an Numerically Controlled Oscillator (NCO). The modulator is implemented using three combinational circuits namely, the phase accumulator, the NCO and the binary to thermometer code converter. These combinational circuits have negative edge triggered registers between them, which are clocked by the *CLK* signal. The combinational circuits are discussed in the next couple of sections.

14.3.4.1 Phase Accumulator and NCO

The NCO is a digital implementation of a sinusoidal oscillator. The advantage of an NCO is that the frequency of the sinusoidal wave produced by the NCO and its phase can be altered in real time by programming the NCO. The basic operation of the NCO is described next. The NCO is implemented as a lookup table (LUT) that stores quantized and rounded values of the sinusoidal wave. The index of the LUT represents the angle for which the sinusoidal value needs to be found. If 2^n is the depth of the LUT where n is the number of bits needed to address the lookup table, then each address of the lookup table stores 2^n equally spaced samples of the sinusoidal wave for an angle of 0° to 360°. The LUT is then addressed by a self-incrementing counter known as the phase accumulator. Thus, when the phase accumulator and the NCO are clocked using a clock signal with a frequency of f_{clk}, the phase accumulator causes evenly spaced values of the sinusoidal wave to be read out from the NCO depending on the value by which it increments. The output frequency generated by the NCO is given by the equation:

$$f_{\text{out}} = \frac{f_{\text{clk}} \Delta\theta}{2^n}, \tag{14.2}$$

where f_{out} is the frequency of the output digital sinusoidal wave generated by the NCO, f_{clk} is the clock signal driving the phase accumulator and the LUT and $\Delta\theta$ is the value by which the phase accumulator increments on every clock cycle. In order to change the frequency produced at the output of the NCO we need to control the phase accumulator increment, namely $\Delta\theta$ based on the value of the binary input signal that needs to be modulated. The depth of the LUT from 14.2 is one of the factors that controls the granularity or resolution with which we can choose output frequencies. The width of each word stored in the LUT also plays a role in finding a sine value with sufficient accuracy. The quality of the output frequency is measured by the spectral purity of the output signal. This is measured by a parameter called the Spurious Free Dynamic Range (SFDR). A good rule of thumb to attain a good SFDR at the output of the NCO is that the SFDR in dB is six times the width of the phase accumulator in bits. For example if we had a phase accumulator that is 9 bits wide, the SFDR would be 54 dB. This is provided the width of the word stored in the LUT is wide enough. However, the word length of the LUT does not improve the SFDR when it becomes wider. An advantage of using an NCO to generate the two FSK tones is that continuous phase is guaranteed at the output of the digital modulator. When the binary input changes from a logical "0" to a logical "1," the frequency of the NCO changes output changes smoothly without giving a kink at the output of the modulator.

One of the optimizations that can be made to the NCO is that the LUT need not store sinusoidal values for all input angles. In fact, the size of the LUT can be reduced by a factor of 4 due to the inherent quarter wave symmetry of the sinusoidal wave. Depending on the quadrant of the input angle, the sine wave can be generated from just a quarter of the samples for a full cycle. A register is required at the output of the phase accumulator since the previous value of the phase accumulator needs to be stored to allow it to increment itself. We choose the NCO to have a phase accumulator that is 9 bits wide and have an output that has a precision of 8 bits. This gives us an SFDR of 54 dB, which is a reasonable amount of rejection for our application. An estimate of the f_{clk} signal made in Sect. 14.3.2 gives us the value 1.4 MHz. In order to transmit wireless data using orthogonal FSK we have the condition that the $f_1 - f_2$ is an integer multiple of the data rate, R_B, which is 32 *kbps*. By Nyquist's theorem the maximum frequency that can be represented without losing information using a clock rate of 1.4 MHz is half its value. By this argument, the values taken by f_1 and f_2 will be less than 700 kHz. But we also need to have a high enough value of f_1 and f_2 so as to make it easy to demodulate at the receiver side. Hence, we choose the phase accumulator increment $\Delta\theta_1$ as 59. This gives us a tone that is less than f_{clk} by close to a factor of 3. This gives the frequency of the first tone from (14.2) to be,

$$f_1 = \frac{f_{\text{clk}} \times 59}{512}. \tag{14.3}$$

We choose the second tone to have a frequency three times less than that of f_1. This is done by choosing the phase accumulator increment, $\Delta\theta_2$, as 117. Also if we choose f_{clk} to be an integral multiple of R_B, then the condition for orthogonal FSK will be satisfied. We can choose f_{clk} to be 40 times R_B so that it is less than the estimated value of 1.4 MHz. In this case $f_1 = 151.04\,kHz$ and $f_2 = 453.12\,kHz$. Note that the values of f_1 and f_2 can be left completely programmable, achieving a Software Defined Radio(SDR) transmitter. But we need additional eight inputs for this, hence this was not done for the sub-threshold IC.

14.3.4.2 Binary to Thermometer Code Converter

This circuit block converts a binary encoded digital signal to a thermometer code. The thermometer code is essentially a one hot code, which has as many LSB "1"s in the code as the unsigned number represented by the binary encoded signal. The use of the Thermometer Code is to pre-process the digital signal before passing along an input to the Digital to Analog Converter (DAC). The higher order bits of the digital signal are converted to thermometer codes while the lower order bits are left binary encoded. Assuming that the binary encoded signal does not change by large values, this will ensure that thermometer code changes by very few bits for small changes in the binary code. Whereas if the binary code is used as input to the DAC, even small increments in value have the potential to change many bits in the code. This causes ripples in the output of the DAC and is undesirable. In our design, we convert four MSBs to thermometer encoded bits and leave the four LSBs as binary encoded bits.

14.3.5 Digital to Analog Converter

The circuit diagram of the DAC is shown in Fig. 14.5. The DAC has a reference current mirror, M1 biased by resistor Rcm. It also has as many current mirrors reflecting the reference as the number of input bits. The input to the DAC is a 19-bit digital signal. The top 15 MSBs are thermometer encoded and the four LSBs are binary encoded. Hence, the DAC has 19 current mirror legs. Figure 14.5 shows two of the current mirror legs of the DAC. The inputs T_i and T_{ib} are the *ith* thermometer

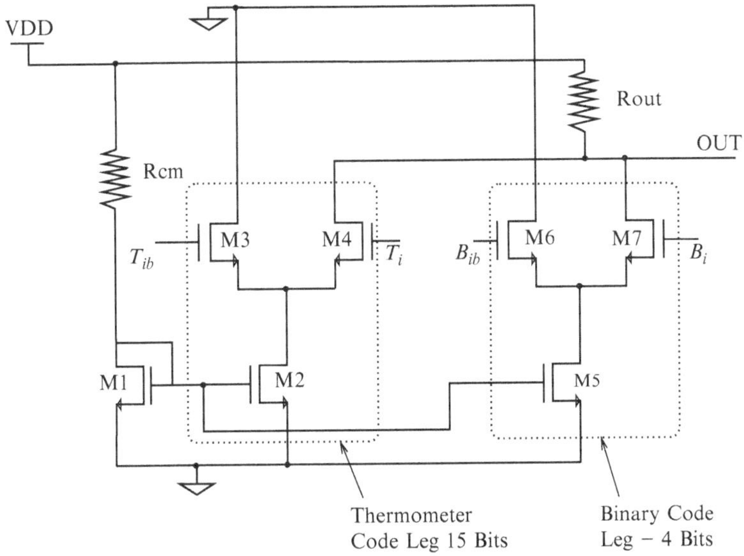

Fig. 14.5 Digital to analog converter

encoded bit and its complement. The inputs B_i and B_{ib} are the *ith* binary encoded bit and its complement. The DAC works by switching the current mirrors ON depending on the value of the input bits and measuring the voltage across the Rout resistor due to this current. The input bits control the NMOS transistors, M3, M4, M6 and M7. For any of these legs, if the input bit is LOW, then the NMOS on the left i.e. M3 or M6 turns ON and prevents the current mirror leg from conducting current. If the input bit is HIGH, then the NMOS on the right turns ON and allows the leg to mirror the current in the reference transistor M1. The difference between the current mirrors for the thermometer code and the binary code is in the size difference between M2 and M5. The *W/L* of M5 used in the current mirrors for the binary encoded bits is 1.3, 2.6, 5.2, 10.4 from LSB to MSB. The *W/L* ratio doubles for every next MSB. The transistors corresponding to M2 have a *W/L* of 20.8 for all the current mirror legs for the 15 thermometer encoded bits. This allows the DAC to modulate the voltage at OUT based on the weighted current flowing through Rout and through different current mirror legs.

14.3.6 Common Source Amplifier

A common source amplifier is needed at the output of the DAC to amplify the signal and drive the antenna. The common source configuration is shown in Fig. 14.6. The common source amplifier is an inverting amplifier. In this configuration, note that there are no bias resistors biasing the gate of the transistor M1. The gate of M1 is

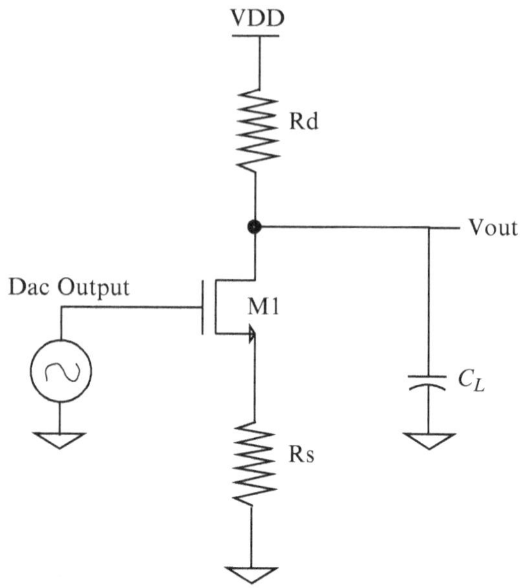

Fig. 14.6 Common source amplifier

connected to the output of the DAC. The gate is thus biased by the DC component of the sinusoidal voltage from the output of the DAC. The amplifier is powered by a very low VDD. Under this condition, other amplifiers such as the source follower or common drain amplifier do not function correctly. The transient response of the common source amplifier will be shown in Sect. 15.5.

14.3.7 Antenna

An on-chip antenna is used to transmit the signal from the amplifier. However, due to the low frequency of operation, the length of the antenna coil needs to be comparable to half the wavelength of the transmitted signal, which is around 300 m. We have used an antenna coil of a length of only 0.2 m due to area constraints on the chip. However, an external antenna can be used to transmit the signal if needed.

14.4 Design Specifications

14.4.1 Link Budget Analysis

The link budget analysis [4] is used in any wireless communication system to calculate the transmit power required at the transmitter side based on certain criteria and assumptions. In this section the link budget analysis is done for a digital non-coherent BFSK transmitter. The design constraints assumed are as follows: the transmit distance is 1,000 m and the data rate, R_B, of the voice signal to be modulated is 32 kbps. The link budget analysis is done as follows.

- *Modulation Technique:* The modulation technique used is FSK. With FSK, two separate frequencies are chosen, one frequency representing a logical "zero," the other representing a logical "one." For non-coherent FSK the channel bandwidth is typically twice the data rate. In our case we have chosen f_1 as 151 kHz and f_2 as 453 kHz as given in Sect. 14.3.4.1. The channel bandwidth is 302 kHz. R_B. This will also aid in easily designing a reliable and robust receiver system as the two transmitted frequencies are wider apart.
- *Noise Floor:* The noise power in watts is given by

$$N = kTB, \tag{14.4}$$

where k is Boltzmann's constant in J/K, T is the system temperature usually assumed to be 290 K, and B is the channel bandwidth in Hz

$$\begin{aligned} N &= 1.38 \times 10^{-23}\,\mathrm{J/K} \times 290\,\mathrm{K} \times 302\,\mathrm{kHz} \\ &= 1.209 \times 10^{-12}\,\mathrm{mW} \\ &= -119.18\,\mathrm{dBm}. \end{aligned}$$

A typical low-cost receiver would add about 15 dB to the noise floor. Hence, the receiver noise floor is -104.18 dBm
- *Receiver Sensitivity:* The required signal strength needs to be determined at the receiver input. For non-coherent digital BFSK modulation using orthogonal signals, the probability of bit error at the receiver is given by the following expression [5].

$$P_\mathrm{b} = \frac{1}{2}\mathrm{e}^{-E_\mathrm{b}/2N_0}. \tag{14.5}$$

By plotting (14.5) we can find the bit energy to noise ratio, E_b/N_0 required at the receiver for a particular Bit Error Rate (BER). An E_b/N_0 of 100 gives us a BER of 10^{-19}. We can calculate the Signal to Noise Ratio (SNR) required at the input of the receiver using the equation:

$$\mathrm{SNR} = \frac{E_\mathrm{b}}{N_0}\frac{R_B}{B}. \tag{14.6}$$

Here R_B is the data rate and B is the channel bandwidth. The SNR required at the receiver input is 12.21 dB. The required signal strength at the receiver or the receiver sensitivity is given by adding the receiver noise floor and the SNR. The power required at the receiver for correct demodulation, P_rx is the receiver noise floor plus the SNR which is –91.97 dB.

- *Path Loss:* The path loss in dB is given by the equation:

$$L = 20\log_{10}\left(\frac{4\pi D}{\lambda}\right), \tag{14.7}$$

where D is the transmit distance, λ is the free space wavelength at the carrier frequency which can be taken as $(f_1 + f_2)/2$. If the carrier frequency is taken as 453 kHz, we get the path loss, L, as 21.98 dB. The higher the carrier frequency used, the more the path loss.
- *Antenna Gain:* The transmitter antenna gain, G_{tx} and the receiver antenna gain, R_{tx} can both be taken as 0 dB. This is a reasonable assumption for a simple dipole antenna.
- *Fade Margin:* Signal fading occurs when waves emitted by the transmitter travel along a different path and interfere destructively with waves traveling on line of sight path. A good rule of thumb for the fade margin is 20 dB.
- *Link Calculation:* The transmit power required, P_{tx} is given by the expression:

$$\begin{aligned} P_{tx} &= P_{rx} - G_{tx} - G_{rx} + L + FadeMargin \\ &= -91.97\,dBm - 0\,dB - 0\,dB + 21.98\,dB + 20\,dB \\ &= -49.99\,dBm. \end{aligned}$$

If we have a safety margin of 49.99 dB then we have to design the chip with a transmit power of 0 dBm or 1 mW. If the output signal has a peak voltage of V_P, and if we assume a 50 Ω resistance on the output node, then the peak voltage required to get a transmit power of 1 mW is given by

$$V_P^2 = 1\,mW \times 50\,\Omega, \tag{14.8}$$

$$V_P = .22V. \tag{14.9}$$

Equation 14.9 needs to be taken into account for the DAC and the amplifier that are going to provide the output signal to the antenna.

14.5 Summary

In this chapter we covered the entire design considerations of the wireless BFSK transmitter chip. We presented the architecture of the chip and analyzed each of the modules separately. We also went through a link budget analysis to determine the amount of transmit power needed to transmit a signal over a distance of 1,000 m.

References

1. HSPICE. www.synopsys.com/products/mixedsignal/hspice/hspice.html (2007)
2. Taiwan Semiconductor Manufacturing Company Ltd. www.tsmc.com (2007)
3. Jayakumar, N., Garg, R., Gamache, B., Khatri, S.: A PLA based Asynchronous Micropipelining Approach for Subthreshold Circuit Design. In: Proc. Design Automation Conference, pp. 419–424 (2006)
4. Proakis, J.: Digital Communications. Boston, McGraw-Hill (2001). http://www.amazon.de/exec/obidos/redirect?tag=citeulike01-21&path=ASIN/0072321113
5. Xiong, F.: Digital Modulation Techniques, Second Edition (Artech House Telecommunications Library). Artech House, Inc., Norwood, MA (2006)

Chapter 15
Implementation of the Chip

15.1 Overview

In this chapter we cover all implementation aspects of the chip. We start with an overview of the design flow used (in Sect. 15.2. Next, in Sect. 15.3, we discuss how we translate the BFSK circuit (written in Verilog) to a netlist (of a network of PLAs). In Sect. 15.4, we discuss how we verify the dynamic compensation circuit through SPICE simulations. The design of the DAC and amplifier circuitry is covered in Sect. 15.5. Some special considerations that need to be taken care of for this chip, including some additions required for the sake of improved testability and improved yield are discussed in Sect. 15.6. In this section, we also discuss how we created separate voltage domains to enable a comparison of the sub-threshold implementation of the BFSK circuit with a regular super-threshold standard-cell-based version. The details of how we implemented the standard-cell-based version of the BFSK circuit are covered in Sect. 15.7. The design of the IO pads and the ESD structures used is covered in Sect. 15.8. In Sect. 15.9, we present how the entire chip was integrated and how we decided the pin-out for the IC. Layout details of the all the components of the IC are covered in Sect. 15.10. We explain how we verified the design before tape-out in Sect. 15.11.

15.2 Design Flow

The steps of the design flow to be used are shown in Fig. 15.1 and briefly described in the remainder of this section.

- First the design specification (obtained by user requirements such as frequency of data being transmitted, available bandwidth, distance of transmission, etc.) was determined.
- Next, the HDL code to implement the specification was developed. VHDL was used for this step.
- This code was synthesized next, resulting in an RTL description of the design.

N. Jayakumar et al., *Minimizing and Exploiting Leakage in VLSI Design*,
DOI 10.1007/978-1-4419-0950-3_15,

Fig. 15.1 Design flow

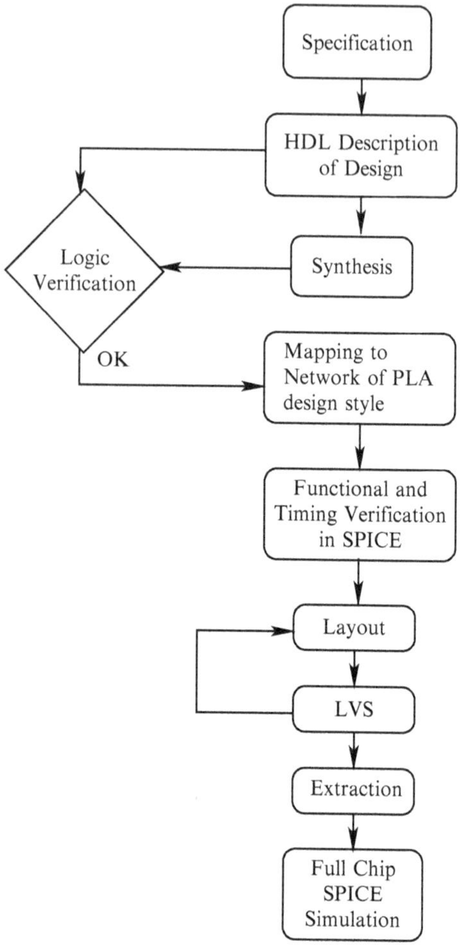

- The synthesized code was verified against the HDL, by running functional test vectors.
- Next the design was mapped to a network of PLA-based design flow. We used the synthesis code from [5] for this purpose. The size of each of the PLAs to be used in the design was determined at this point based on the number of PLAs required for the design (area) and the speed of operation of the PLAs (latency and throughput). At the end of this step, a SPICE level netlist description of the design is obtained.
- A functional and timing verification is done on the SPICE level schematic. This simulation is done across all process corners. This validates and tests the design of the circuit to some extent. The design of the circuit can be changed based on the results of this step.

- Using the net list of PLAs, which results from the previous step, the layout of each PLA was drawn using the TSMC 0.25-μm process. Additionally, the layout of IO pads, ESD cells, and analog components was also drawn.
- Layout Versus Schematic (LVS) verification was performed next to ensure that there were no layout errors.
- Finally, the design parasitics were extracted, and the entire design was simulated in SPICE as a final sign-off.

15.3 HDL to Netlist Flow

The HDL description of the digital portion of the circuit was written using VHDL The external inputs and outputs of the digital BFSK modulator are described in Table 15.3. The output of the binary to thermometer code converter block is a 19bit wide digital signal. These 19 signals are fed into the input of the DAC and cannot be viewed externally.

The HDL description was then synthesized using a synthesis tool for an FPGA. The synthesis tool used was Xilinx ISE Foundation [3]. The synthesis tool output is a gate level description of the implemented circuit. This description is then converted into a logic format for further synthesis optimization using the multi-level logic synthesis tool SIS [6]. Using SIS, the blif file representation of the digital modulator circuit is then mapped into a network of PLAs. The algorithm used for this mapping is given in [5]. The algorithm involves the following steps. First, a technology-independent optimization is done on the given multi-level circuit. Next this circuit is decomposed into a network of nodes with each node having at most five nodes. Then these nodes are then levelized, meaning that each node is assigned a level that is one larger than the largest level of all its fanin nodes. The next step in the algorithm is to group nodes together and fit them in a PLA of the given maximum size. We use folded PLAs to fit more logic in a PLA compared to a non-folded PLA. Folded PLAs are explained in [5], and in our case, we fold only inputs. The logic representation of the multi-level network of PLAs that is obtained in this step is then used to create a SPICE netlist description of the digital modulator circuit. The SPICE netlist is used as a golden schematic netlist, for LVS verification purposes as well. All the PLAs used to build the circuit have the same size so that they have approximately the same delay. Also this makes the layout of the PLAs easier as the footprint of the metal wires is same for all PLAs and only the transistors in the PLAs are modified based on the logic implemented. In order to find the size of the PLA to be used, we did the following experiment.

We used a set of circuits from the *mcnc91* benchmark circuits, where each circuit was decomposed into a multilevel network of PLAs using the PLA decomposition algorithm, for several PLA sizes. Depending on the number of PLAs and the number of levels in the multilevel circuit and the delays of the PLAs, we found that PLAs

Table 15.1 PLA configuration

PLA (In,Out,Cube)	T_{pchg}	T_{eval}	Total no. of PLAs	No. of PLA levels for NCO block	Delay	Throughput
(8,6,12)	45 ns	35 ns	4+24+3	19	710 ns	1.4 MHz

with sizes of 8-12 inputs, 4-6 outputs and 12-18 rows have a low delay as well as a small area of implementation.

The size of the PLA we use for this circuit is 8 input, 6 outputs, 12 cubes. From SPICE simulations the evaluation and precharge periods of a PLA of this size for the TSMC 0.25 μm process were found to be: $T_{eval} = 35\,ns$ and $T_{pchg} = 45\,ns$.

Each of the three logic blocks that constitute the BFSK modulator shown in Fig. 14.2 are implemented using combinational logic. The combinational logic is implemented using a multi-level network of PLAs. The NCO block has the largest delay as it requires much more logic than the other two blocks. It also has more number of levels of PLAs than the other two blocks. Table 15.1 shows the maximum throughput that we can attain using this particular PLA size.

The output of this step is a logical description of the network of PLAs used to implement the digital BFSK modulator. From this logical description a SPICE schematic is created. The next step in the implementation process is to interface the digital circuitry with the dynamic delay compensation circuit described in Sect. 14.3.3.

15.4 SPICE Verification of Dynamic Compensation

The dynamic delay compensation circuit is interfaced with the digital BFSK modulator circuit. An initial simulation is shown in Fig. 15.2. In this case we have configured the beat clock signal to speed up the PLAs. The signal "nandout" in this figure represents the *pull-up* signal shown in Fig. 10.4. This instructs the phase detector and charge-pump circuit shown in Fig. 10.3 to pull-up the bulk node. Whenever there is a low going pulse on the "nandout" signal, we see that the bulk node called "bulkn" in Fig. 15.2 is pulled up. However the "bulkn" node that represents the body terminal of of the NMOS transistors in the design is very noisy with a ripple close to 100mV on every clock cycle. Notice that this ripple is not caused during the downward going pulse of the "nandout" signal and is not due to the charge-pump circuit.

From Fig. 15.2, it can be seen that during the precharge period when the "clk" signal is low, the bulk node gets pulled up and during the evaluation period when the "clk" signal is high, the bulk node gets pulled down. The reason behind this effect can be explained using Fig. 10.1. Notice from this figure that each PLA has a large parasitic drain bulk capacitance due to transistors in the PLA connected to the *dummy wordline*. During every precharge phase, the *dummy wordline* is pulled up

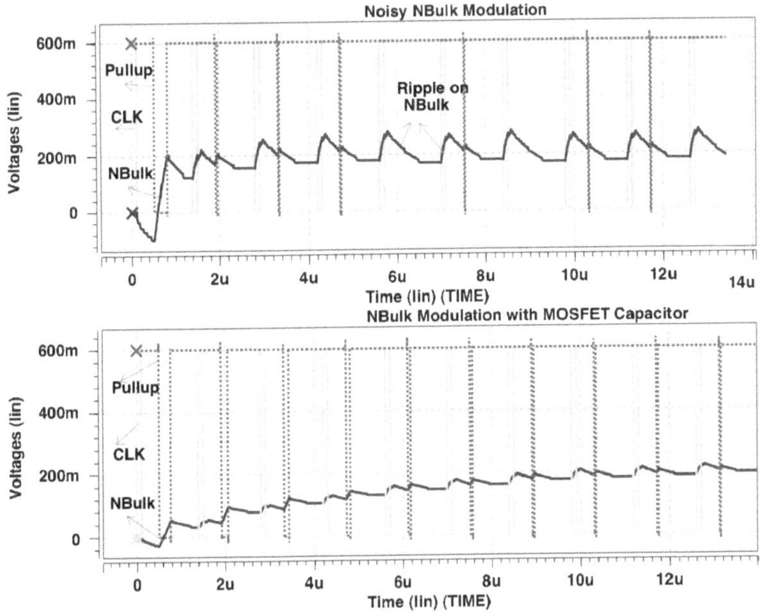

Fig. 15.2 Dynamic bulk node modulation

to VDD and during every evaluation period, the *dummy wordline* is pulled down to GND. This transition couples into the Nbulk node, making it noisy. In order to fix this problem, we have added a capacitor to the bulk node of the NMOS transistors to filter out the noise. The charge-pump devices are made wider so that they can overcome the effect of this capacitor. The capacitor is realized using a MOSFET transistor's gate terminal, with the drain, source and body terminals connected to GND. This is a non-linear capacitor varying from 100 to 180 pF for a bulk node voltage swing of 0 to 0.5 V. The lower part of Fig. 15.2 shows the modulation on the bulk node after adding the MOSFET capacitor. Now the ripple on the bulk node is only 25 mV. We also ran SPICE simulations in which the objective was to slow down the PLAs by configuring the *beat clock (BCLK)* signal as shown in Fig. 10.5. These simulations were run across all corners provided by TSMC for their process.

15.5 DAC and Amplifier Design

The DAC and Amplifier driving the antenna are using the circuit diagrams shown in Sect. 14.3.5 and Sect. 14.3.6, respectively. The following steps are followed to design the DAC and the amplifier.

- The resistors Rcm and Rout of the DAC are designed to be surface mounted resistors outside the chip. This allow us to tune these resistors in real time to enhance the output signal. Two external pins in the pin-out of the chip are reserved for these two resistors.
- The resistors Rs and Rd of the Amplifier are also designed as surface-mounted off-chip resistors. Hence, the Amplifier is also connected to two external pins.
- The output of the amplifier is connected to an on-chip coil antenna. The capacitance of the antenna was estimated by finding the capacitance of a small segment of the antenna structure using Space3d [2] and extrapolating that value for the entire antenna. The total capacitance of the antenna was estimated at around 80pF.
- The output voltages of the DAC and the Amplifier need to have a peak voltage value in accordance with the value calculated in (14.9).
- Sample waveforms at the output of the DAC and the Amplifier are shown in Figs. 15.3 and 15.4, respectively. The output of the amplifier was loaded by an 80 pF capacitor. The output of the DAC and Amplifier are shown alternating between the two frequency tones.

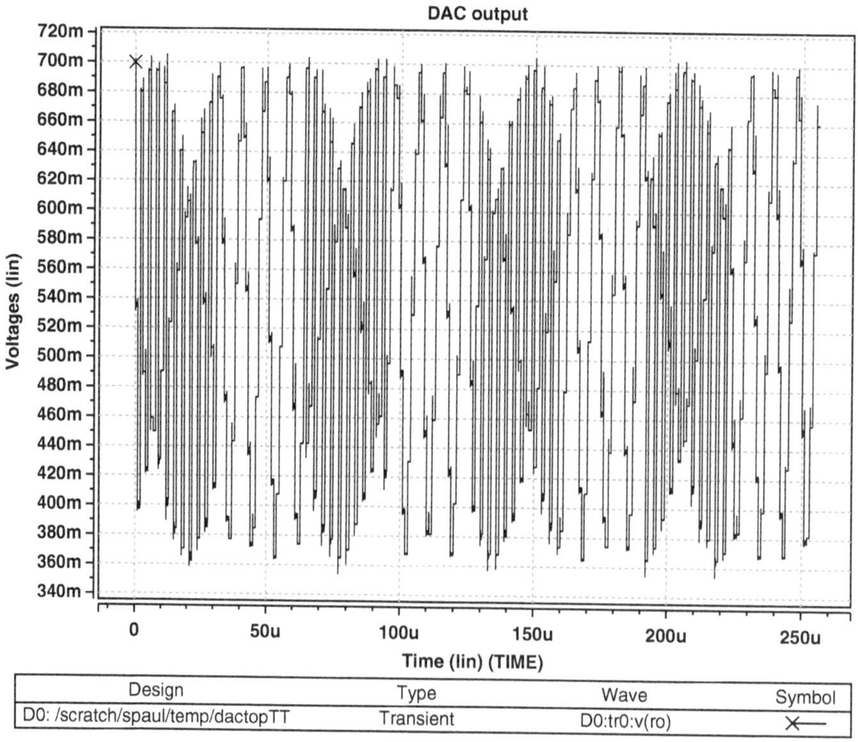

Fig. 15.3 DAC output

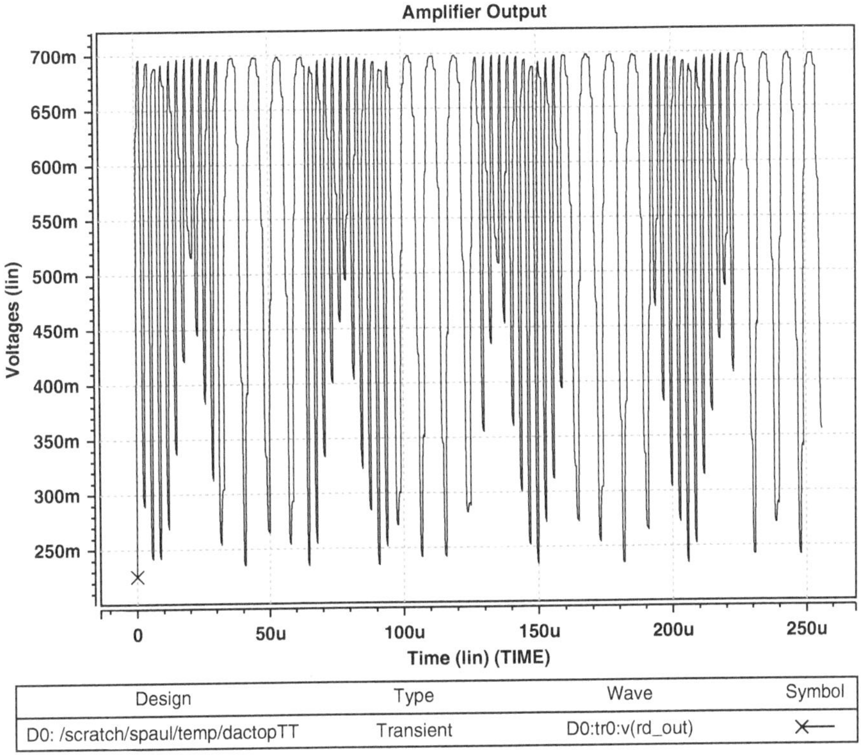

Fig. 15.4 Amplifier output

15.6 Special Considerations

15.6.1 Testability and Redundancy

Various testability features were built into the design. The use of these features is to test each component of the chip individually to verify functionality. They also serve as a backup against failure of one of the components. The following are the testability features that are incorporated in the design.

- A standalone PLA is included in the design along with the other PLA components that make up the digital modulator circuit. The PLA is designed in such a way that the two outputs of the PLA toggle continuously when the clock waveform is applied. The result of this test verifies the functionality of the PLAs, which are the basic building blocks in the design.
- The 8-bit output of the NCO block is directly sent to eight I/O pads on the chip. These pads are bi-directional. This means that these pads on the chip can either be used to get the digital 8-bit sine wave value from the output of the NCO, or can be used as an 8-bit input to the binary to thermometer code converter. This

feature is important since it takes into account the scenario in which only one of the digital modulator or the DAC is functionally correct. In this scenario, these bi-directional pins may be used to excite the correctly functioning blocks in the design.

- The output of the DAC can be measured using an oscilloscope, at the pin that connects the external DAC drive resistor R_{out} to the chip. This allows the DAC to be tuned and tested individually based on its output waveform. This gives us the option of directly using the DAC with an external amplifier and antenna.
- The output of the common source amplifier also can be scoped externally using the pin connected to the R_D resistor. This signal may also modulate an off-chip antenna, instead of the on-chip antenna.
- The output of the amplifier is connected to the antenna through a pass gate that is controlled by a signal called *Anton*. This signal is used to disconnect the on-chip coil antenna by turning off the pass gate if needed.

15.6.2 Voltage Domains

One of the objectives of this experiment is to compare the operation of a sub-threshold circuit with a standard cell-based implementation. The two circuit realizations operate at different VDD values. In order to isolate these two implementations, we need one extra voltage domain for the standard cell implementation. A voltage value of 2.5 V is used for this domain, since 2.5 V is the nominal operating voltage for the TSMC 0.25 μm process. For the targeted process, we have specified the sub-threshold design to work at a VDD of 0.6 V. The inputs to the sub-threshold digital modulator circuit cannot be on the same voltage domain. This is because designing I/O drivers at such a low voltage is not reliable. Hence, we use another voltage domain (higher than 0.6 V) so that the inputs to the sub-threshold circuit operate at this higher voltage. We chose the *VDD* of this domain to be 1 V. One of the built-in testability features of this chip is that the outputs of the sub-threshold digital modulator circuit, if needed, can be sent directly off-chip to an external DAC and antenna. We however found that there was no off-the-shelf DAC that had an input voltage rating of less than 2 V. Hence, the outputs of the sub-threshold circuit needed to be driven to a voltage value of at least 2 V. Hence, another voltage domain with a VDD of 2 V was used.

We thus have four separate VDD domains on the chip. All these domains have a common *GND* to make the power distribution easier. The following special conditions need to be addressed when we have signals that cross two different voltage domains.

- A higher voltage signal cannot drive a pass gate of a lower voltage domain. In this case we buffer the signal with a buffer operating on the VDD of the lower voltage domain before driving the pass gate.
- A higher voltage signal can drive the gate of a transistor in a lower voltage domain.

- To buffer a signal from lower voltage to higher voltage domain, we use custom-designed level shifters.

15.7 Standard Cell-Based BFSK Design

We also implemented a traditional Standard Cell-based BFSK design on the chip for a head-to-head comparison with the sub-threshold approach. The design flow for the standard cell portion of the design consisted of the following.

- We used the same HDL code used for the sub-threshold design.
- The synthesized HDL code was mapped into a library of standard cells that consisted of various inverters (2×, 12×, 36× 108×), and NAND gates(2-input, 3-input).
- The standard cell design is not connected to a DAC and an antenna.
- The mapped design was then placed and routed using the SEDSM [1] from Cadence.
- The inputs to the Standard cell design are *64kinstd*, *Clkstd*, *Resetstd*,
- The output of the Standard cell design is an 8-bit vector *Stdout*, which represents the 8-bit output of the NCO.

15.8 IO Pad and ESD Diode Design

The circuit diagram of a general Pad cell with ESD diodes is shown in Fig. 15.5. The transistors MP1 and MN1 are the primary ESD diodes. The transistors MP2 and MN2 represents the inverter driving an internal signal towards the pad to an

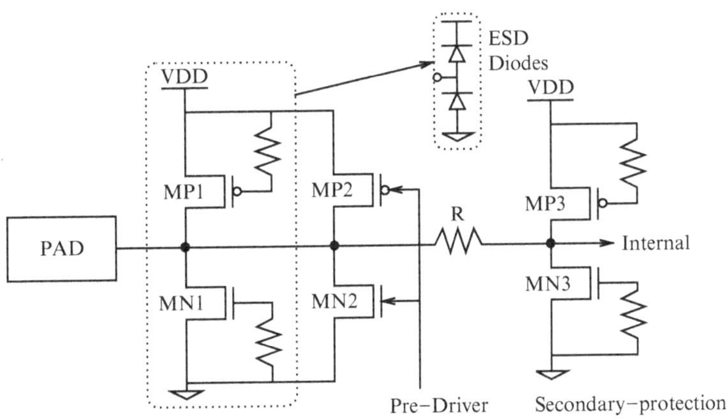

Fig. 15.5 PAD cell schematic

off-chip component. MP3 and MN3 are ESD devices giving further protection. The resistance R has a value of approximately 200-Ω.

We have used four separate voltage domains on the chip. Because of this the pads used for the signals can be classified as follows.

- *Power Supply Pad.* These pads do not have any I/O drivers. They have the ESD diodes shown in Fig. 15.5 and are used for the VDD (for all domains) and GND signals.
- *Digital Input Pad.* These pads have ESD diodes with input drivers driving the external signal towards the chip.
- *Digital Output Pad.* These pads have ESD diodes with output drivers driving the internal signal towards the pad.
- *Digital I/O Pad.* Along with ESD diodes, these pads have both input and output drivers. The output drivers are tristated when this pad is receiving an input signal.
- *Analog Signal Pad:* The analog signals do not have any I/O drivers. Some analog signals do not have the ESD diode connected to VDD. This constraint is used when the peak value of the analog signal can take a higher value than the VDD connected to the ESD diode.

15.9 Chip Integration and Pin-out

The integration of the chip mainly involves deciding the number of pins on the chip. The pin-out for the standard cell implementation of the BFSK transmitter is shown in Table 15.2. The pin-out for the sub-threshold implementation is shown in Table 15.3. We need 80 pins. Note that pins 80, 1, 20, 21, 40, 41, 60, 61 are dummy pins and these are at the corners of each side of the chip. Some of the sensitive signals are shielded using static signals and/or supply signals. An estimate of

Table 15.2 Chip pin-out: standard cell BFSK portion

	Pin Number	Description
Pin Name	Domain 4, VDD = 2.5V	
39	GND	Ground
42	VDD	Supply
43	GND	Ground
44	Resetstd	Active high, Reset signal for Std Cell BFSK
45	32kinstd	Binary input signal for Std Cell BFSK
46	Clkstd	Clock signal for Std Cell BFSK
47–49	Stdout$< 8 : 6 >$	Digital output of Std Cell BFSK
50	VDD	Supply
51	Anton	Active high, loads the Amplifier with on-chip antenna
52	GND	Ground
53–57	Stdout$< 5 : 1 >$	Digital output of Std Cell BFSK
58	GND	Ground
59	VDD	Supply

Table 15.3 Chip pin-out: Sub-threshold BFSK portion

Pin number	Pin name	Description
	Domain 1, $VDD = 1V$	
7	Dacin	Active high, apply external DAC input to pins 23–25,28–30,33–34
8	Clk	Clock signal to BFSK modulator, shielded by static signals
9	VDD	Supply
10	Reset	Active high, resets the BFSK modulator output
11	GND	Ground
12	32kin	Binary input to modulator
13	sdrouten	Active high, NCO output sent to pins 23–25,28–30,33–34
14	VDD	Supply
15	Beat Clk	Reference clock for dynamic compensation, Shielded by VDD
16	VDD	Supply
17–18	GND	Ground
	Domain 2, $VDD = 2V$	
19	GND	Ground
22	VDD	Supply
23–25	In2vOut2v$< 1 : 3 >$	NCO output or DAC input
26	GND	Ground
27	VDD	Supply
28–30	In2vOut2v$< 4 : 6 >$	NCO output or DAC input
31	GND	Ground
32	VDD	Supply
33–34	In2vOut2v$< 7 : 8 >$	NCO output or DAC input
35	Testplaout1	Pla test signal 1
36	GND	Ground
37	VDD	Supply
38	Testplaout2	Pla test signal 2
	Domain 3, $VDD = 0.6V$	
62	GND	Ground
63	VDD	Supply
64	AmpRdRes	Drain resistance of amplifier, shielded
65	GND	Ground
66	VDD	Supply
67	AmpRsRes	Source resistance of amplifier, shielded
68	GND	Ground
69	VDD	Supply
70	DacCmRes	DAC current mirror resistance, shielded
71	GND	Ground
72	VDD	Supply
73	DacDriveRes	DAC output resistance, shielded
74	GND	Ground
75	VDD	Supply

(continued)

Table 15.3 (continued)

	Pin name	Description
Pin number	Domain 1, $VDD = 1V$	
76	GND	Ground
77	Bulkinout	Monitor or Force NBulk node
78	PdKickSupply	Supply voltage of charge pump
79	VDD	Supply
2	VDD	Supply
3	GND	Ground
4	VDD	Supply
5	GND	Ground
6	VDD	Supply

the floorplan of the chip is made and signals are buffered depending on the distance that they have to travel. A SPICE level schematic of the entire chip can thus be constructed by including, the digital modulator, the DAC, amplifier and connecting their input, output signals to pad cells. The antenna is represented by a large capacitor.

15.10 Layout

The layout of the PLA block used in the design is shown in Fig. 15.6.

Each of the PLAs have the same number of inputs, outputs and cubes. The logic implemented by the PLAs however is different. The transistors connected to the bitlines, wordlines and output lines need to be changed for each of the PLAs depending on the function implemented. The layout of the DAC and the amplifier are also done. The transistor lengths used for these analog components are three times the minimum length. This increases the variation tolerance of these components. The antenna is implemented as a coil. The antenna is made of five metal layers, as well as the poly layer. The metal layers and poly layer are all connected to each other by contacts. The pad cells are laid out in accordance with the design rules associated with pads and ESD cells from TSMC. Guard rings are used to prevent latch-up in the ESD diodes. The resistor R, seen in Fig. 15.5 is realized using N-type diffusion material to have a resistance of around 200 Ω.

The vacant areas in the chip are then filled with metal to satisfy the fill rules of the design process. These metal fills are wired up to act as a decoupling capacitance between supply and ground nodes. This serves to drastically reduce supply voltage noise.

The standard cell layout is done using the SEDSM tool [1]. This layout is merged with the rest of the components to get the entire die layout shown in Fig. 15.7.

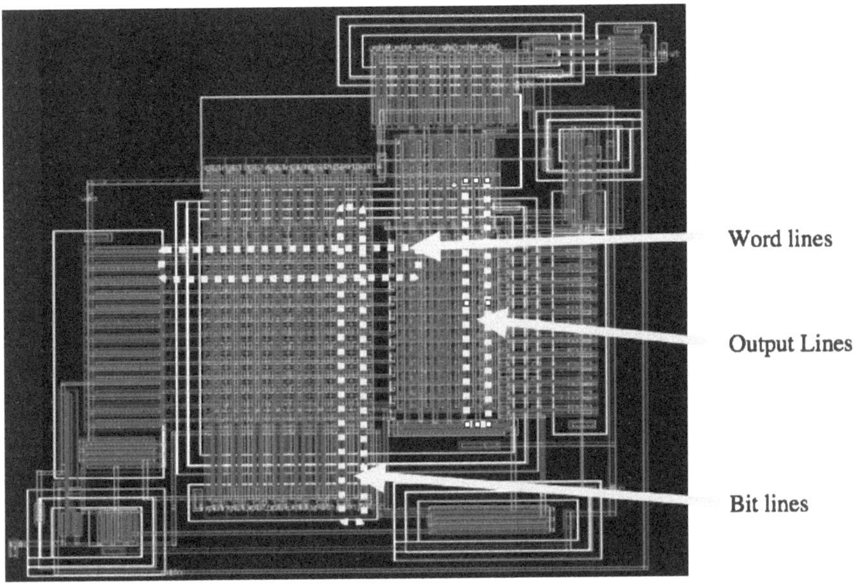

Fig. 15.6 PLA layout

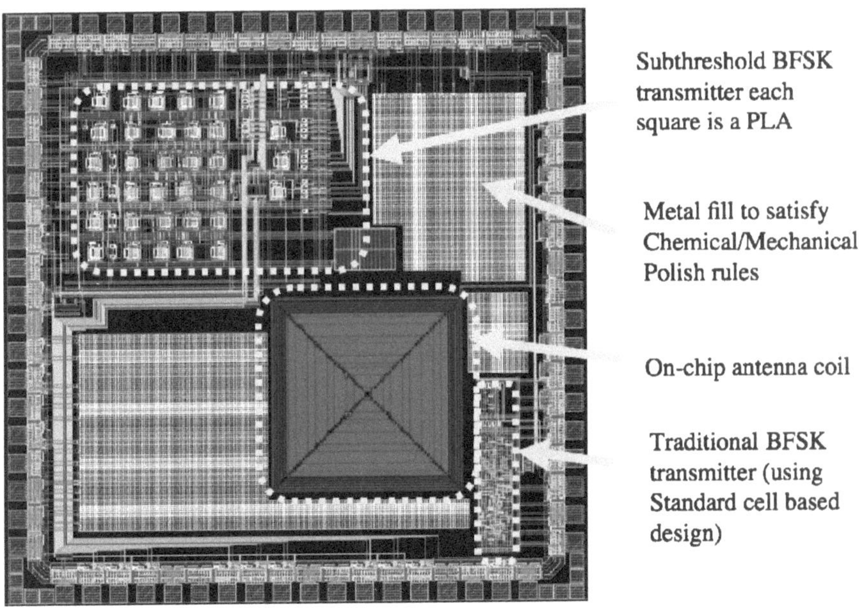

Fig. 15.7 Die Layout

15.11 Summary of Verification Methodologies

The following verification methodologies were used at various stages during the design flow shown in Fig. 15.1.

- *Combinational Verification.* This is a verification step done after synthesis. The logical representation of the circuit after optimization is functionally verified against the initial HDL description.
- *SPICE Verification.* SPICE-based verification is done after mapping the logic netlist into a multi-level network of PLAs. SPICE verification is done to verify functional correctness as well as correctness of dynamic bulk node modulating compensation circuit.
- *LVS.* A layout vs. schematic step is performed after layout designing to verify the correctness of the layout. This was performed using the ASSURA LVS tool [4].
- *RC Extraction and Verification.* An RC Extraction of the chip is performed after the LVS step. This populates the circuit schematic with various parasitic resistors and capacitors. A SPICE level simulation of this extracted netlist is required to verify that the circuit behavior has not been adversely affected by parasitics. The SPICE level simulation also covers the bulk node modulation by the compensation circuit. This is important as there may be extra parasitic capacitances on the Nbulk node, which would require stronger devices on the charge-pump device.

15.12 Summary

In this Chapter we went over the implementation details of the wireless BFSK transmitter. The design flow of the chip was discussed and we explained each step in the flow. The chip was divided into four different voltage domains to isolate the standard cell implementation, and provide higher VDD for inputs and outputs to the sub-threshold circuit. The steps taken for the layout were also discussed.

References

1. Cadence Design Systems, Inc., 555 River Oaks Parkway, San Jose, CA 95134, USA: Envisia Silicon Ensemble Place-and-route Reference (1999)
2. van Genderen, A.J., van der Meijs, N.P.: Space3d Capacitance Extraction User's Manual. Delft Univ. of Technology, Delft, The Netherlands (1997)
3. Xilinx Inc.: ISE Foundation. http://www.xilinx.com/ise/logic_design_prod/foundation.htm (2007)
4. Design Systems Inc., C.: ASSURA Layout vs. Schematic Verifier. http://www.cadence.com/products/dfm/assura_lvs (2007)

5. Khatri, S.P.: Cross-talk Noise Immune VLSI Design using Regular Layout Fabrics. Ph.D. thesis, University of California, Berkeley (1999)
6. Sentovich, E.M., Singh, K.J., Lavagno, L., Moon, C., Murgai, R., Saldanha, A., Savoj, H., Stephan, P.R., Brayton, R.K., Sangiovanni-Vincentelli, A.L.: SIS: A System for Sequential Circuit Synthesis. Tech. Rep. UCB/ERL M92/41, erl, University of California, Berkeley, CA 94720 (1992)

Chapter 16
Experimental Results

16.1 Overview

The tests carried out to verify functionality of the chip (die photo in Fig. 16.1) are first presented in Sect. 16.2. Next, in Sect. 16.3, we present the tests on the dynamic compensation circuit of the sub-threshold portion of the chip. The operating range of the chip is explored in Sect. 16.4. In Sect. 16.5, we show the FFT of the output of the DAC and the output of the amplifier on the chip. A comparison of the power consumption and performance of the sub-threshold implementation and the standard-cell implementation of the BFSK circuit is covered in Sect. 16.6.

16.2 Functional Verification

The VDD domains 1 and 4, which correspond to the sub-threshold BFSK inputs, and DAC and amplifier outputs are powered ON. The *reset* signal is held LOW. The DAC and Amplifier are biased using resistances determined during the circuit design phase. The output of the DAC for an input signal that makes a LOW to HIGH transition is shown in Fig. 16.2.

Note that the DAC output clearly shows two tones depending on the value of the input.

16.3 Dynamic Compensation Circuit

The dynamic compensation circuit stabilizes circuit delay by modulating the bulk node of the NMOS transistors in the design as explained in Sect. 14.3.3. Figure 16.3 shows an oscilloscope plot of the bulk node voltage and power supply of the sub-threshold circuit. Here the external beat clock has been fixed to a particular delay. Notice that when the supply voltage that is the bottom signal in the plot fluctuates from its nominal value, the bulk node voltage that is the top signal in the plot is

N. Jayakumar et al., *Minimizing and Exploiting Leakage in VLSI Design*,
DOI 10.1007/978-1-4419-0950-3_16, © Springer Science+Business Media, LLC 2010

Fig. 16.1 Die photo

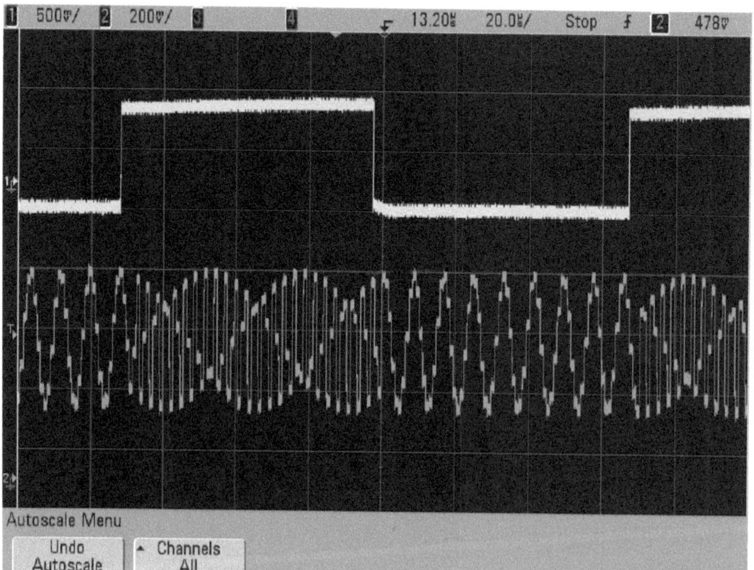

Fig. 16.2 BFSK modulation

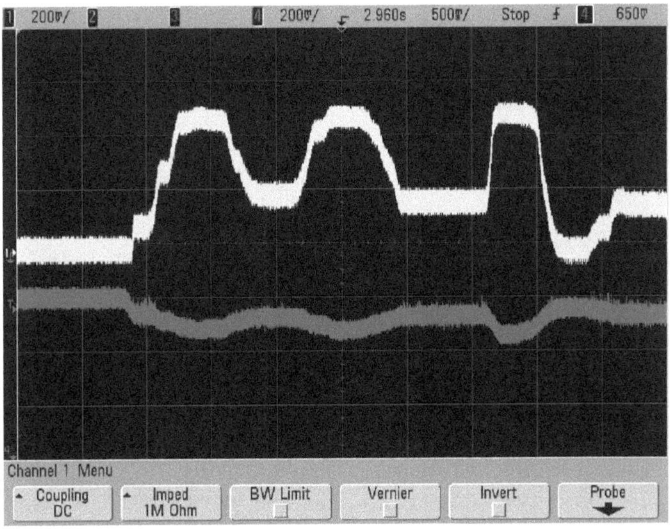

Fig. 16.3 Bulk node voltage modulation with VDD

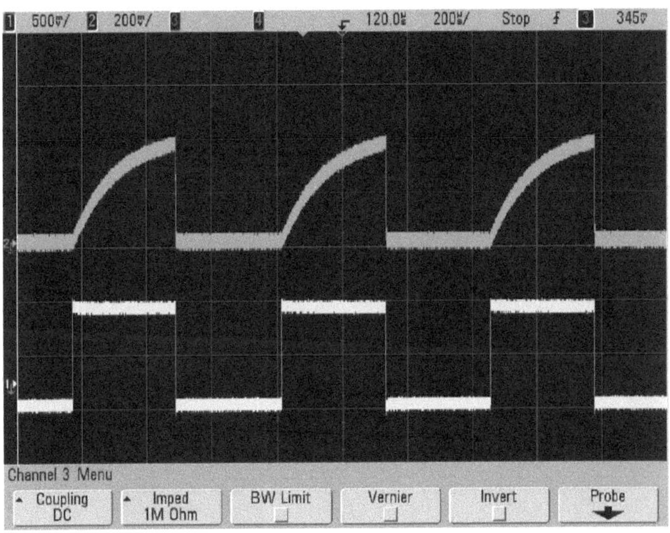

Fig. 16.4 Bulk node voltage modulation with *BeatClock*

immediately modulated in the opposite direction to compensate the circuit delay with respect to power supply variation. Thus, the reference circuit delay is kept in phase with the external reference signal.

Figure 16.4 plots the bulk node voltage in the top half and the external beat clock signal in the bottom half. Here the beat clock is held high for several clock cycles and then held low for several clock cycles. When the beat clock signal is held high the charge pump forward biases the bulk node and the circuit speeds up. When the beat clock signal is held low the bulk node is driven low and the circuit slows down. The bulk node is clearly modulated up and down when the phase of the beat clock signal changes verifying the operation of the dynamic body bias circuit with respect to the external reference signal.

16.4 Operating Ranges

The supply voltage for the digital BFSK modulator circuit was varied from 0.4 V to 0.62 V. The maximum frequency of operation at these voltages was determined by observing the output of the source amplifier. When the frequency is too high, the sine wave at the output of the amplifier gets distorted. The maximum operating frequencies over a set of supply voltages is plotted in Fig. 16.5.

This figure shows two curves that correspond to a bulk node voltage value of 0 V and 0.45 V, respectively. This plot shows the range of frequencies over which the dynamic compensation circuit can track the reference beat clock. Notice that the maximum speed of operation increases quadratically as the supply voltage increases.

The power consumed by the circuit at these operating voltages and frequencies is shown in Fig. 16.6. The power consumed is plotted for the maximum and minimum voltage value that the bulkn node can take. The power consumed is the product of the average current flowing through the digital BFSK modulator voltage source. Note that a different voltage source is used for the DAC and the amplifier.

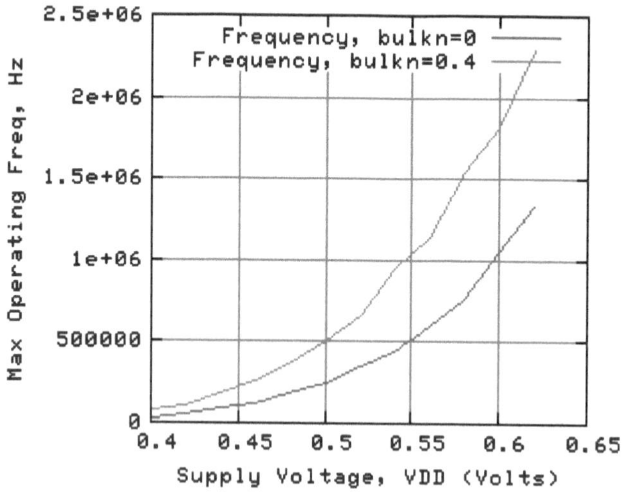

Fig. 16.5 Maximum operating frequencies

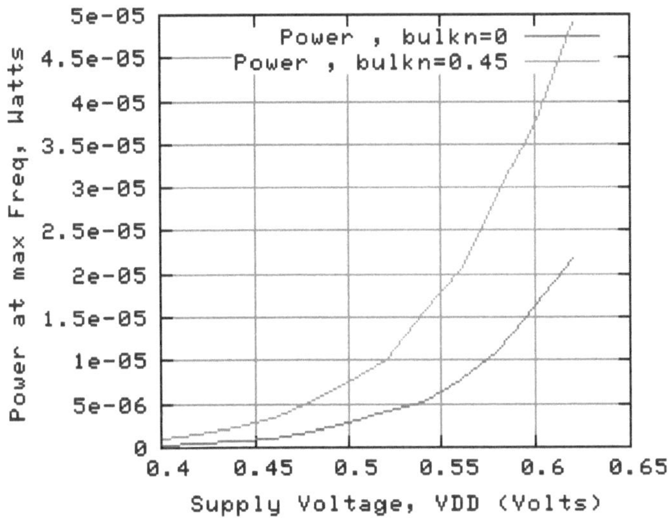

Fig. 16.6 Power consumed at maximum operating frequency

16.5 Spectrum of Output Sinusoidal Signals

The Fast Fourier Transform (FFT) of the output of the DAC is shown in Fig. 16.7. Here the input bitstream is continually alternating between a logical "zero" and a logical "one" at a frequency of 32.25 kHz. The clock frequency, f_{clk}, of the sub-threshold circuit is set at 1 MHz, which is an integer multiple of the input bit rate. From the FFT we see the two transmitted tones at 113 and 342 kHz, respectively.

Similarly Fig. 16.8 shows the FFT of the output of the amplifier for the same signal when the amplifier is loaded by the on-chip antenna coil. Notice that the secondary unwanted peak between the two tones is around -11 dB below the fundamental tone. Also through Matlab simulations we found that a signal with a spectrum that has the secondary unwanted peak at -10 dB was demodulated correctly at the receiver side. This simulation was done for the worst-case noise and attenuation considered in the link budget analysis in Sect. 14.4.1. The receiver architecture used was a standard receiver for demodulating non-coherent BFSK signals [1].

16.6 Comparison with Standard Cells

The power consumed by the sub-threshold BFSK Modulator was compared with the power consumed by the standard cell BFSK implementation. This is shown in Table 16.1. From this table we see that the power consumed by the standard cell-based circuit implementation is 19.4× more. The standard cell-based design is specified to operate at a supply voltage of 2.5V. Note that the standard cell-based

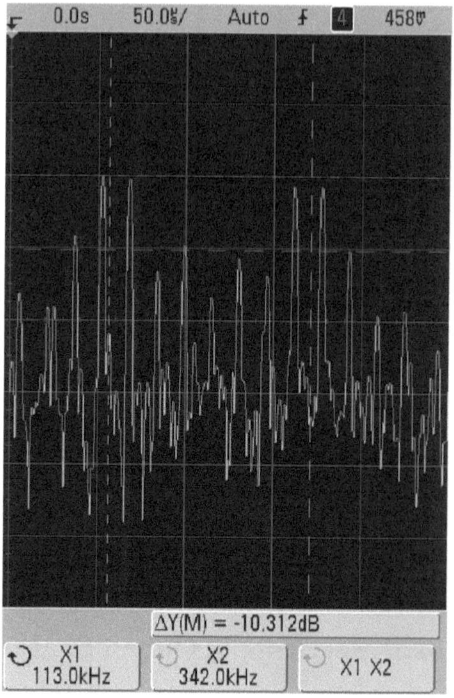

Fig. 16.7 FFT of DAC output

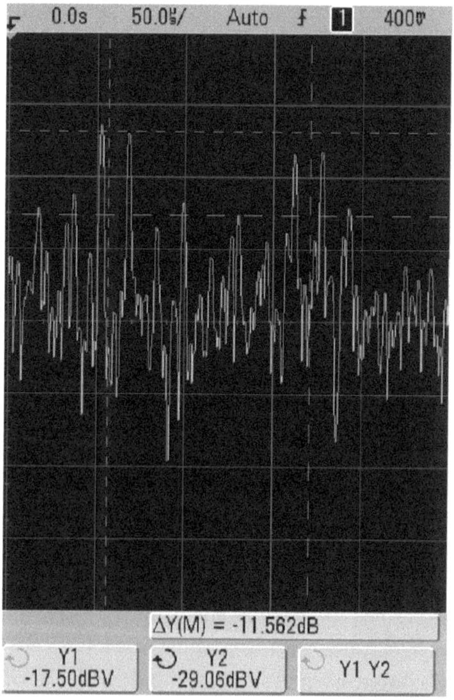

Fig. 16.8 FFT of amplifier output

Table 16.1 Sub-threshold vs. standard cell power consumption

Design style	VDD	Clock frequency (MHz)	Average current (μA)	Power dissipation (μW)
Sub-threshold	0.6	1.05	44.7	26.8
Standard cell	2.5	1.05	208.0	520.0

design is capable of operating at higher speeds. The standard cell design does not have any compensation scheme that compensates circuit delay for PVT variations, which are higher when operating near the sub-threshold region. Hence, it would not function correctly under varying operating conditions. Because of this we do not compare the standard cell-based design power at a lower voltage of operation.

16.7 Summary

In this Chapter, we presented results from the fabricated wireless BFSK transmitter chip. We verified the functionality of the digital BFSK circuit and the dynamic delay compensation circuitry. We also analyzed the spectrum of the output signal and showed that the transmitted signal spectrum can be suitably demodulated with a standard non-coherent receiver architecture. We also showed that the power consumed by a standard cell-based implementation of the same circuit on the same die is $19.4\times$ more.

Reference

1. Xiong, F.: Digital Modulation Techniques, Second Edition (Artech House Telecommunications Library). Artech House, Inc., Norwood, MA (2006)

Summary and Future Work

Power Consumption in VLSI circuits is a critical issue in the semiconductor industry today. For many applications such as portable devices, low power consumption is a first-order design constraint. Several of these applications need extreme low power but do not have high-speed design requirements. In these cases Sub-threshold circuit design techniques can be used to provide extreme low power solutions, by sacrificing some of the circuit performance. The problem with sub-threshold circuits however is that these circuits exhibit an exponential sensitivity to process, voltage and temperature (PVT) variations.

In this part of the book we have presented implementation details of a robust sub-threshold design flow, which uses circuit level PVT compensation to stabilize circuit performance. This involves compensating the delay of a circuit over PVT variations by using an external reference clock. The compensating circuitry modulates the bulk node of transistors in the circuit depending on the phase difference between the circuit delay and the reference clock signal. The circuit is implemented using a Network of PLAs in which all PLAs are of the same size. Therefore, each PLA has the same delay, and this ensures that the critical path delay to be compensated is the same across the entire circuit.

We have designed and fabricated a sub-threshold wireless BFSK transmitter chip using this robust sub-threshold design methodology. The chip is capable of broadcasting a signal over a distance of 1,000 m. For comparison purposes we have also implemented a BFSK transmitter using a traditional standard cell flow on the same die and shown that the sub-threshold approach consumes 19.4× lower power than a traditional standard cell-based implementation.

Future work includes constructing an antenna for wireless transmission and constructing a receiver that can be used to demodulate the signal transmitted by the BFSK transmitter. This can be used to test and verify the distance over which the wireless transmitter can operate. Also the speed of the sub-threshold circuit can be improved drastically by using heavily pipelined circuits.

Conclusion

In this book we have presented various techniques to deal with the problem of increasing leakage in modern VLSI designs. These have included techniques to minimize as well as exploit leakage currents.

Part I of this book focused on leakage reduction techniques. In chap. 2, we first presented a survey of the state- of-the-art leakage minimization techniques in use today. We then described new leakage minimization techniques we invented in the next few chapters. In Chaps. 3 and 4, we presented different algorithms to compute the Minimum Leakage Vector (MLV). The algorithm presented in Chap. 3 was an Algebraic Decision Diagram (ADD) based algorithm to generate a histogram of leakage current over all input vectors of a circuit. The algorithm in Chap. 4 was an algorithm that used signal probabilities to guide the search for an optimal MLV. This algorithm was augmented to use statistical leakage variation information to find an optimal MLV that reduced the mean and standard deviation of leakage. In Chap. 5 we described a new low-leakage standard cell-based ASIC design methodology that we call the HL methodology. The HL methodology is a type of power gating methodology that selectively uses high-V_T PMOS (header) or NMOS (footer) transistors within the standard cells and thus reduces leakage to a low and precisely estimable value. One of the key advantages of this technique is the fact that no nodes float during standby. In Chap. 6, we presented a technique that combined input vector control and circuit modification to enable leakage reduction without any delay penalty. In Chap. 7, we presented a scheme to dynamically find the optimum reverse biasing voltage. There is no one magic bullet that can mitigate the leakage power problems IC designers face today. Each of the leakage reduction techniques presented in Part I have their own advantages and disadvantages and the decision on which technique is used is dependent on the intended application and the investment in time and money that a designer is willing to make.

In Part II of this book, we looked at leakage currents from a completely different viewpoint and sought to exploit leakage currents rather than minimize them through the use of sub-threshold circuit design. We first explored the opportunities that sub-threshold circuits offered in Chap. 9 and also revealed the factors that are preventing sub-threshold circuit design from becoming mainstream. In next few chapters, we detailed our methodologies and flows to help make sub-threshold circuit design feasible and practical. In Chap. 10 we presented an adaptive body biasing scheme to combat the high sensitivity of sub-threshold circuits to process, voltage and temperature (PVT) variations. In Chap. 11, we then presented a study on what determines the optimum voltage of a circuit from a minimum energy point of view. We performed this study for a dynamic network-of-PLA style of design – the design style we chose to implement digital sub-threshold circuits. In Chap. 12, we presented an asynchronous micropipelining technique, to help claw back the delay penalty associated with sub-threshold circuit design (and ultra-low voltage design in general).

While Part two of this book detailed some design methodologies and circuit techniques for sub-threshold circuit design, in Part three of the book we presented how we designed and tested a test application – a sub-threshold wireless BFSK transmitter IC to provide silicon validation of our sub-threshold circuit design techniques. In Chap. 14, we covered the design constraints and the architecture of the design. The implementation details were covered in Chap. 15 while Chap. 16 went over some data collected from experiments on the fabricated IC. The data collected showed that our sub-threshold implementation consumed 19.4× lower power than a super-threshold standard-cell-based implementation. Thus we proved the feasibility of our sub-threshold design approach.

It is hoped that this book inspires a researcher to do further research into low power design. For a designer, we hope that book provides solutions to some of the power problems an IC designer may face.

Index

Kolibri 360 GmbH
Ferdinand-Jühlke-Straße 7
99095 Erfurt, Deutschland
produktsicherheit@kolibri360.de